Water Regulations 1999

Course & Assessment

Introductory module

Introduction to the course

Water Byelaws have been a fact of life for plumbers as long as most of us can remember. On 1 July 1999 local Water Byelaws were replaced, by the Water Supply (Water Fittings) Regulations 1999.

This learning package has been developed by BPEC to help plumbers update their knowledge of this important legislation which is used to control the installation and use of water supply systems. The package will help you to implement Water Regulations competently, not only in new installation work, but also to ensure you are able to recognise and rectify contraventions that you come across in existing water supply systems.

Aimed primarily at practising plumbers who already have a knowledge of hot and cold water systems, the package will be of interest, and be useful, to water inspectors and anyone who is involved with the installation, design, or specification of water systems or with buildings containing water systems. It will also be of benefit to those learners with limited previous knowledge.

The package will serve – firstly, as a means of learning, and
– secondly, as a reference for future use.

The aims of this package are to enable course participants to:

- become familiar with the contents of the Water Supply (Water Fittings) Regulations 1999;
- generally understand what the Regulations mean, and recognise the role of the Regulations in preventing waste, misuse, undue consumption, contamination and erroneous measurement of water;
- know how to comply with the Regulations in practice;
- be aware of the primary legislation underpinning the Regulations;
- have an understanding of the documents used to enhance and explain the meaning of the Regulations and in particular:
  - the Regulators' Specification on the Prevention of Backflow;
  - the WRAS Water Regulations Guide;
- bring you up to a standard where you have the option of attending a BPEC Training and Assessment Centre for a One Day Training and Assessment Course.

The package

The package contains this introductory booklet, followed by 12 learning modules and finally one support document in the form of a glossary of terms.

The learning modules are supported by a video outlining the contents of the Regulations which can be found at http://bpec.org.uk/shop/free-learning-resources/

Contents

About this course *(general introduction to the course)*

How to tackle the work

The way in which you learn is largely up to you, but the following notes will give some guidance to those of you who feel they need it.

You should first look at the video which will take about half an hour and provide an overview of the Regulations.

Don't try to do too much at once. It is quite a deep subject, and it's probably best that you take it in easy stages, one step at a time and one module at a time.

BPEC Training and Assessment Centres will send you the Training Package at least two weeks before the Training and Assessment Course if you have chosen this route.

1. *Read the requirements of the Regulations and try to understand what they say. These are given at the start of each learning module, or in the case of Module 2 at intervals as you go through. In places the meanings will be quite plain, whilst other parts will take more consideration and thought.*

2. *Read the explanations given within the modules which should be easier to understand than the actual Regulations. This is because there are examples given of how the Regulations are applied.*

3. *Go through it more than once if you need to. The more times you look at it, the better it will stick in your memory.*

4. *Use the self-assessment questions to confirm whether you have taken it all in. These are spaced out in the modules and also serve to break the reading and give you something to stimulate your thinking. Write your answers in pencil not ink so any errors can be easily put right.*

5. *Check your answers against the model answers at the back of each learning module.*

Repeat this procedure throughout the package, and good luck in your endeavours.

What next?

You will now have a thorough knowledge of the Water Supply (Water Fittings) Regulations 1999, and your Training Package will provide you with an excellent source of reference.

You can leave it at that or you can consider attending a BPEC Training and Assessment Centre, where you can be assessed, and where upon successful completion BPEC will issue a BPEC Water Regulations Certificate.

BPEC Training and Assessment Centres

BPEC has approved a number of Training and Assessment Centres throughout the UK at which a one day training and assessment will be carried out, covering the Water Supply (Water Fitting) Regulations 1999.

Training

Training at the Centres will be delivered by the Centre first of all sending the Training Package to you at least three to four weeks before the one day course. You will have to work your way through all the questions in the twelve modules before you attend the course. (You must take your Training Package to the Centre on the training and assessment day.)

The Centre will give you a further half day training, followed by a half day of assessment.

Assessment

The assessment will consist of one multi-choice question paper for which you will be able to refer to your Training Package manual or notes taken on the day.

The multi-choice questions will consist of a question and four answers, one correct and three wrong answers. You simply have to select the correct answer. The time allowed is 75 minutes.

You are required to achieve an 80% pass rate in both Section A and Section B to be successful. You have to achieve 70% or above in your first attempt. Candidates who achieve less than 70% in the first attempt in one or both sections will not have successfully completed the assessment.

Candidates who achieve 70% or more in the first attempt have a second attempt to achieve 80% on both sections. Candidates who do not achieve 80% in their second attempt will not have successfully completed the assessment.

Successful candidates will receive a BPEC Water Regulations Certificate.

Assessment only

Some Centres may make arrangements for candidates who have purchased the BPEC Water Regulations Training Package to attend the Centre for Assessment only.

Why do I need a certificate?

Under the Water Supply (Water Fittings) Regulations 1999, Water Companies are encouraged to keep a list of Approved Contractors (sometimes called Approved Plumbers) for the area they cover. Holding a BPEC Certificate should prove that your knowledge of the Water Supply (Water Fittings) Regulations 1999 is good enough for you to become an Approved Contractor provided you meet the other criteria laid down by the Water Companies.

What are the advantages of becoming an approved contractor?

Approved contractors do not need to notify certain installations which non-approved contractors must notify. Water Companies will compile a list of approved contractors, which the general public will use, and it is possible that Local Authorities and Housing Associations will insist that plumbing contractors are approved contractors. Approved contractors will be permitted to self-certify their work, thus reducing the need for constant monitoring by Water Inspectors.

Last step

Should you want to attend a BPEC Training and Assessment Centre on a One Day Training and Assessment Course, where a successful pass will bring you a BPEC Water Regulations Certificate, please phone 01332 376000 for further details.

Further reading/references

It is not absolutely necessary to purchase any other books, as the package is self contained and aimed to ensure that learning will take place without the additional burden of further book purchases. However, you may feel the need to read more deeply as you progress, or you may wish to go into the Regulations in more depth after you have completed this course.

Whichever way you look at it, the following book list could be of some use.

Editor:	Water Regulations Advisory Scheme (WRAS)
Date:	2000
Title:	Water Regulations Guide
Publisher:	Water Regulations Advisory Scheme (WRAS)
	WRc Evaluation and Testing Centre Ltd., 13 Willow Road, Pen-y-Fan Industrial Estate, Crumlin, Gwent, NP11 4EG

Editor:	Water Regulations Advisory Scheme (WRAS)
Date:	Published 6 monthly, 1 April and 1 October
Title:	'Water Fittings and Materials Directory'
Publisher:	Water Regulations Advisory Scheme (WRAS)
	WRc Evaluation and Testing Centre Ltd., 13 Willow Road, Pen-y-Fan Industrial Estate, Crumlin, Gwent, NP11 4EG

Editor:	DWI – Water Supply (Water Quality)
Date:	June 2016
Title:	The Water Supply (Water Quality) Regulations 2016
Publisher:	HMSO

Editor:	DWI – Private Water Supplies
Date:	June 2016
Title:	The Private Water Supplies (England) Regulations 2016
Publisher:	HMSO

Editor:	DWI – The Water Act
Date:	2014
Title:	Water Act 2014 Chapter 21
Publisher:	HMSO

Authors: HSE
Date: 2014
Title: Legionnaires disease Part 2: The control of Legionella bacteria in hot and cold water systems
Publisher: Health and Safety Executive

Authors: BSI
Date: 2011
Title: BS 8558, Guide to the design, installation, testing and maintenance of services supplying water
              for domestic use within buildings and their curtilage – Complementary guidance to BS EN 806
Publisher: BSI 389 Chiswick High Road, London, W4 4AL

Authors: BSI
Date: 806-1:2000, 806-2:2005, 806-3:2006, 806-4:2010, 806-5:2012
Title: Specifications for installations inside buildings conveying water for human consumption
Publisher: BSI Standards Ltd 2012
              389 Chiswick High Road, London, W4 4AL

Water Industry Act 1991:

Water Supply (Water Fittings) Regulations 1999

An Open Learning Course

Module 1

Introduction, background and legislation

Introduction and background to the Regulations

The control of water supply installations in England and Wales has been radically revised by the introduction of the **Water Supply (Water Fitting) Regulations 1999**.

Following a period of deliberation and consultation, the **Secretary of State for the Department of the Environment, Transport and the Regions (DETR) used his powers under Sections 73, 74, 75, 84 and 213(2) of the Water Industry Act 1991(a) to make Water Regulations** to control the installation and use of water fittings. This resulted in the making of the **Water Supply (Water Fitting) Regulations 1999 which came into force on 1 July 1999**.

It should be noted that these **Regulations apply only in England and Wales**. Similar provisions have been implemented by the Scottish Office and are being implemented by the Northern Ireland Office in their areas of jurisdiction.

We have in this country, a long history of Water Byelaws, administered and enforced by local water suppliers. As long ago as 1823 the Manchester and Salford Act was brought about to *'prevent the wilful and negligent use of water'*, a practice in water wastage control which spread to other private and municipal water companies throughout the country.

The Water Act of 1945 formalised this arrangement by placing an obligation on water undertakers to enforce water byelaws *'for preventing waste, undue consumption, misuse or contamination of water'*. From 1945, Water Byelaws were required to be based on the Government's 'model' and needed the approval of the Government Minister with responsibility for water supply before they could be implemented. Generally, Byelaws were made to expire after a life of 10 years, after which time they were renewed and updated as became necessary.

This constant renewal of Water Byelaws continued until the latest Byelaws, based on the 1986 'model', were replaced by new Water Regulations on **1 July 1999**.

So! How do regulations differ from Byelaws?

Quite simply, **Byelaws were made locally** and applied only in the area in which they were made. For instance, Byelaws made by Thames Water applied only within the Thames Water area of supply. Thames Water were held responsible for enforcement of their own Water Byelaws.

Water Regulations on the other hand, **are National Regulations**, made by the Department of the Environment, Food and Rural Affairs (DEFRA) **and they apply to every installation in England and Wales that is supplied from a public main by a Water Undertaker**.

The responsibility for enforcement of the Regulations is placed on the Water Undertakers.

The new Regulations have similar aims to previous Byelaws, but a new way of implementing them. They are made **'as a means of preventing waste, undue consumption, misuse, contamination, and the erroneous measurement of water'**.

Whilst the aims are similar, as you go through this package you will find that there are quite a number changes in the way water fittings have to be installed and used.

What legislation is in place to control water installations in this country?

The main legislation governing the making of **Water Regulations is the Water Industry Act 1991** and **Sections 73, 74, 75, 84, and 213(2)** in particular are relevant.

- Section 73   Offences of contaminating, wasting and misusing water etc, (legal action)
- Section 74   Regulations for preventing contamination, waste etc and with respect to water fittings
- Section 75   Power to prevent damage and to take steps to prevent contamination, waste etc
- Section 84   Local authority rights of entry etc
- Section 213(2)   Powers to make regulations.

Extracts from Section 74 are quoted below and main points highlighted:

74-(1) *The Secretary of State may by regulations make such provision as he considers appropriate for any of the following purposes, that is to say:*

(a) *for securing:*

    (ii) **that water in a water main** *or other pipe of a water undertaker* **is not contaminated,** *and*

    (iii) **that its quality and suitability** *for particular purpose* **is not prejudiced, by the return of any substance** *from any premises* **to that main or pipe;**

(b) *for securing* **that water** *which in any pipe connected with any such main or other pipe or which has been* **supplied to any premises by a water undertaker is not contaminated, and that its quality and suitability** *for particular purposes* **is not prejudiced,** *before it is used;*

(c) **for preventing the waste, undue consumption, and misuse of any water** *at any time after it has left the pipes of a water undertaker for the purpose of being* **supplied by that undertaker to any premises;** *and*

(d) *for securing* **that water fittings installed and used** *by persons to whom water is or is to be supplied by a water undertaker* **are safe and do not cause or contribute to the erroneous measurement of any water or the reverberation of any pipes.**

So! By briefly looking at the wording of Section 74, we can see that the Water Regulations made under the Act have been made for the following purposes:

- to make sure that **water is not contaminated, and its quality and suitability for purpose is not prejudiced,**
- to **prevent waste, undue consumption, and misuse of water** supplied by the undertaker, and
- to make sure that **water fittings installed and used are safe and do not cause or lead to erroneous measurement, or reverberation** (vibration/noise) in pipes.

In other words, the Regulations have been written to protect the water supply and to protect users against their own actions.

The Water Supply (Water Fittings) Regulations 1999

These are made in a similar format to Building Regulations and over the years prior to the Water Regulations being made, there was considerable discussion as to whether they should actually be included within Building Regulations and whether they should be enforced by Building Control Officers. After much consideration it was decided that the Water Regulations should remain separate from Building Regulations and continue to be administered and enforced by the water undertakers (suppliers).

What is in the Regulations that I should know about?

Well! You should know, or at least be aware of, pretty well everything that is in the Regulations. However, before getting into detail, let's start with a brief overview which is as follows.

The Water Supply (Water Fittings) Regulations 1999 consists of 14 Regulations which are divided into three parts and supported by three schedules.

Part I 'Preliminary' gives the date at which the Regulations came into force, it gives some interpretations to support and help us to understand the Regulations. Part I also makes statements as to how the Regulations should be applied. **Regulation 1** is supported by **Schedule 1** which describes the **five fluid categories**. (See Module 8)

Part II 'Requirements' sets out what is expected of persons installing water fittings, how water fittings should be installed and used to prevent waste or contamination and places conditions on the materials and fittings that may be used. Part II also requires **contractors to notify water suppliers of certain installations** and encourages the introduction of Approved Contractors Schemes.

Also in Part II, **Regulation 4(3)** is supported by **Schedule 2 'Requirements for Water Fittings'** which deals with the more practical aspects of the Regulations.

Part III Enforcement, as the title suggests, **deals with** aspects of **enforcement, penalties** for contravention of the Regulations, and **dispute procedures**.

Schedule 1 to the Regulations sets out fluid risk categories related to the backflow requirements of Schedule 2.

Schedule 2 to Regulation 4(3) 'Requirements for Water Fittings' consists of 31 separate requirements and really contains the 'meat' of the document, looking at the use of water fittings in all its many aspects. In fact with the exception of the first two modules, the whole of this training package deals with the requirements set out in the Schedule. In Modules 3 to 11 we will be looking at requirements for:

– **Materials** (in Module 3)
– **Requirements for water fittings** (in Module 4)
– **Design and installation** (in Module 5)
– **Commissioning** (in Module 6)
– **Prevention of cross connections** (in Module 7)

- **Backflow** (in Module 8)
- **Cold water services** (in Module 9)
- **Hot water services** (in Module 10)
- **WCs, flushing devices and urinals** (in Module 11) and
- **Sanitary appliances and water for outside use** (in Module 12)

Schedule 3 'Byelaws Revoked' supports Regulation 14 and simply lists the byelaws of the various water undertakings in England and Wales which have been taken off the Statute book and replaced with the Water Industry (Water Fittings) Regulations 1999.

Approved (Guidance) Document

Water Regulations, compared to previous byelaws are less prescriptive, and contain little in the way of technical detail other than that given in Schedule 2 'Requirements to Regulation 4(3) mentioned above.

The Department of the Environment, Food and Rural Affairs (DEFRA) has produced a **Guidance Document** to accompany the Regulations that can be used in a similar way to that of the approved documents that go with Building Regulations.

In addition to the DETR Guidance, the Water Regulations Advisory Scheme (WRAS) which from 1st April 2021 separated into two separate businesses.
1. Water Regs UK promotes compliance with the water fittings regulations and Byelaws and is responsible for operating the Water Industry Approved Plumber Scheme (WIAPS). The company has a new website www.waterregsuk.co.uk
2. Water Regulations Approval Scheme (WRAS) operates a voluntary certification scheme for plumbing products and materials. The company website is: www.wrasapprovals.co.uk One way for installers to check compliance might be to look for products certified by WRAS, Kiwa or NSF, but be aware there may be restriction how and where they are installed.

The use of the approved document is not mandatory and failure to comply with its recommendations cannot in itself lead to any liability under Regulations. However, where an installer can show that his installation is in compliance with the recommendations of the approved document, his installation can be deemed to satisfy the requirements of the Regulations.

It should be remembered also that in law **the burden of proof is with the installer**.

This means it is up to you, the installer, to show proof that your installation complies with the requirements of the Regulations.

In turn, this means you are expected to be competent in what you do. You need to have the qualifications, knowledge, and practical experience to carry out your work in a proper and **'workmanlike manner'**. (Regulation 4(5))

It is the aim of this package to help you to gain a good knowledge of Water Regulations and thus improve your competence in the installation work that you do.

So! Now some work for you. Please turn to the next page for your first set of self-assessment questions.

Self-assessment questions

1. The Secretary of State for the Department of the Environment, Transport and the Regions (DETR) used his powers under Sections 73, 74, 75, and 84 of the Water Industry Act 1991 to make the Water Industry (Water Fittings) Regulations 1999.

   Give the date on which the Regulations came into force

   _____

2. Name the TWO United Kingdom countries that are directly affected by these Regulations.

   _____ *and* _____

3. In the past we have been used to water byelaws. State briefly how the Regulations differ in status from byelaws.

   Byelaws _____

   _____

   Water Regulations _____

   _____

   _____

4. State who is responsible for enforcing the new Water Regulations.

   _____

5. In addition to the Water Regulations document the DEFRA have also produced an Approved (or guidance) Document. We are obliged by law to comply with the Water Regulations. Explain briefly the status of the Approved Document.

   _____

   _____

   _____

Check your answers on page 8.

Summary of main points

Background and legislation

The Secretary of State for the Department of the Environment, Food and Rural Affairs (DEFRA) used his powers under **Sections 73, 74, 75, 84 and 213(2) of the Water Industry Act 1991 to make the Water Supply (Water Fittings) Regulations 1999.**

The Regulations came into force on **1 July 1999 and apply only in England and Wales.**

Similar provisions, based on the Regulations for England and Wales, are made for Scotland. In Scotland Revised Water Byelaws are retained, and Northern Ireland continue with their own Regulations

Our previous history of water byelaws in England and Wales has come to an end. **Byelaws** which were made locally and applied only in the area in which they were made **are now revoked and replaced by the Water Supply (Water Fittings) Regulations 1999.**

The Water Regulations are National Regulations, and apply to every installation in England and Wales that is supplied from a public main by a Water Undertaker.

The Regulations are made **for securing that water in any water main or pipe** supplied by the water undertaker **is not contaminated and that its quality and suitability for purpose is not prejudiced.** They are made for **preventing the waste, undue consumption and misuse of water supplied by the undertaker, and for securing that water supplied by the undertaker is safe and does not contribute to the erroneous measurement of water or reverberation in pipes.**

The responsibility for enforcement of the Regulations is placed on the Water Undertakers.

Water Regs UK promotes compliance with the water fittings regulations and Byelaws and is responsible for operating the Water Industry Approved Plumber Scheme (WIAPS). The company has a new website www.waterregsuk.co.uk

Answers to self-assessment questions

1. The Secretary of State for the Department of the Environment, Transport and the Regions (DETR) used his powers under Sections 73, 74, 75, and 84 of the Water Industry Act 1991 to make the Water Supply (Water Fittings) Regulations 1999.

   Give the date on which the Regulations came into force.

   1 July 1999.

2. Name the TWO United Kingdom countries that are directly affected by the Water Regulations

   The Regulations apply only in

   England and Wales.

3. In the past we have been used to water byelaws. State briefly how the Regulations differ from byelaws.

   Byelaws were made by individual water undertakers and based on a national 'model'. They applied only in the water undertaker's water supply area.

   Water Regulations are National Regulations, and apply to every installation in England and Wales that is supplied from a public main by a Water Undertaker.

4. State who is responsible for enforcing the new Water Regulations.

   The responsibility for enforcement is placed on the Water Undertakers.

5. In addition to the Water Regulations document the DEFRA have produced an Approved (or guidance) Document. We are obliged by law to comply with the Water regulations. Explain briefly the status of the Approved Document.

   The approved document is written to provide practical guidance as to how compliance with the Regulations can be achieved.

What to do next

Well done!

You have made a good start

That is the introduction over with, and you now have a good idea how to tackle these self learning modules.

Now it is time to take a detailed look at the Regulations.

Please go on to **Module 2** and good luck with your studies.

Water Industry Act 1991:

Water Supply (Water Fittings) Regulations 1999

An Open Learning Course

Module 2

The Regulations

Introduction

As you are now aware the correct title for the new Water Regulations is the **Water Supply (Water Fittings) Regulations 1999**. However, the full title is rather a long one so throughout the rest of this package we shall for the most part shorten it.

Where in the future you see the terms Water Regulations, or just simply 'the Regulations', you will know that this means the Water Supply (Water Fittings) Regulations 1999.

We have already given a brief outline of the Regulations in Module 1, so without further ado let's get down to it.

Shortly after publication of the Regulations, certain anomalies became apparent, with the result that an amendment was produced. We now have three documents:

– Water Supply (Water Fittings) Regulations 1999 – Statutory Instrument No.1148

– Water Supply (Water Fittings) (Amendments) Regulations 1999 – Statutory Instrument No.1506

– Additional amendment to The Water Supply (Water Fittings) Regulations 1999 due to EU Exit

The amendments are not extensive, so in this training package the amendments have been incorporated into the text as if they had been accounted for in the first place.

What do you, the plumber, need to know about the Regulations?

Firstly you should read through the Regulations document and familiarise yourself a little with its contents.

Once you have done that we can begin to look more closely at them.

We will take it a stage at a time, and indeed one Regulation at a time, with questions as we go along.

Part I – Preliminary

Part I – Preliminary means just what it says. It is there to help set the scene for what is to come.

Let's begin by looking at Regulation 1.

What does Regulation 1 say?

Regulation 1

Citation, commencement and interpretation

1.-(1) These regulations may be cited as the Water Supply (Water Fittings) Regulations 1999 and shall come into force on 1st July 1999.

(2) In these regulations:

"the Act" means the Water Industry Act 1991;

"approved contractor" means a person who:

(a) has been approved by the water undertaker for the area where a water fitting is installed or used, or

(b) has been certified as an approved contractor by an organisation specified in writing by the regulator;

"the Directive" means Council Directive 89/106/EEC on the approximation of laws, regulations and the administrative provisions of the member States relation to construction Products (b);

"fluid category" means a category of fluid described in Schedule 1 to these regulations;

"material change of use" means a change in the purpose for which, or the circumstances in which, premises are used, such that after that change the premises are used (where previously they were not so used):

(i) as a dwelling;

(ii) as an institution;

(iii) as a public building; or

(iv) for the purpose of the storage or use of substances which if mixed with water result in a fluid which is classified as either fluid category 4 or 5;

"regulator" means:

(a) in relation to any water undertakers whose area of appointment is wholly or mainly in Wales and their area of appointment, The National Assembly for Wales;

(b) in relation to all other water undertakers and their area of appointment, the Secretary of State;

"supply pipe" means so much of any service pipe that is not vested in the undertaker;

and Paragraph 1 of Schedule 2 has effect for the purposes of that Schedule.

So! What does it all mean?

Let's have a look!

As you can see **Regulation 1.-(1)** begins by simply giving the date of commencement of the Regulations.

They came into force on **1 July 1999**.

Regulation 1.-(2) gives a number of **definitions that support the Regulations** and aim to help us to understand what exactly some of the terms mean. Some of these are quite easy to understand, whilst others perhaps need some additional explanation.

Let's have a look at some of them:

Approved contractor means a person who:

(a) has been approved by the water undertaker for the area where a water fitting is installed or used, or

(b) has been certified as an approved contractor by an organisation specified in writing by the regulator.

Reference is made to approved contractors in Regulations 5, 6 and 7, where the Water Supply Industry is encouraged to set up Approved Contractors Schemes. More of this when we get to Section 2 of the Regulations.

Fluid category means a category of fluid described in Schedule 1 of the Regulations. Fluid categories are described more fully in Module 8.

Material change of use means a change in the purpose for which, or the circumstances in which, premises are used. For example a house (dwelling) that is turned into a nursing home (institution) or a church (public building) that is turned into a dwelling. In both cases there are changes in use that affect the requirements of the regulations.

Under 'material change of use' four types of premises are listed (but not defined) Definitions for these are given below:

i) **a dwelling** is a place where people live e.g. house, flat, bungalow etc.

ii) **an institution** is a building that provides living accommodation for, or for the treatment or care of people suffering from illness or disability or those who are unable to care for themselves. Examples include certain hospitals, schools, homes for the young or old, but not day centres.

iii) **a public building** can be described as premises designed and built for use by the general public including such buildings as theatres, schools or colleges of education, public libraries, halls where people meet, and places of worship.

iv) **buildings used for the storage or use of substances of fluid category 4 or 5.** This can include any type of building providing the said substances are stored. (Fluid categories are described more fully in Modules 7 and 8).

Note: There are other terms such as 'premises' which are not defined here but many definitions are to found in other legal documents. This, and a number of other definitions are given for you in a glossary of terms towards the end of this package.

What does Regulation 2 say?

Regulation 2

Application of the regulations

2.-(1) *Subject to the following provisions of this Regulation, these regulations apply to any water fitting installed or used, or to be installed or used, in premises to which water is or is to be supplied by a water undertaker.*

(2) *These regulations do not apply to a water fitting installed or used, or to be installed or used, in connection with water supplied for purposes other than domestic or food production purposes, provided that:*

(a) *the water is metered;*

(b) *the supply of the water is for a period not exceeding one month, or with the written consent of the undertaker, three months; and*

(c) *no water can return through the meter to any pipe vested in a water undertaker.*

(3) *Except for the purposes of Paragraph 14 of Schedule 2 (prevention of cross connection to unwholesome water), these Regulations do not apply to water fittings that are not connected or to be connected to water supplied by the water undertaker.*

(4) *Nothing in these Regulations shall require any person to remove, replace, alter, disconnect or cease to use any water fitting that was lawfully installed or used, or capable of being used, before 1st July 1999.*

Regulation 2 tells us what the Regulations apply to, and then goes on to say what they do not apply to.

So! What do the Regulations apply to?

It's quite straightforward really! **Regulation 2.-(1)** says in effect that:

- **the Regulations apply to all water fittings installed or used** after 1 July 1999;
- **in ALL premises that are supplied by the water undertaker;**
- they apply to **new works**, and **alterations to existing work**, including any **connections and disconnections**;
- they apply to all **permanent installations** and to **some temporary installations**.

It should be noted here that the Water Regulations apply to all water fittings **installed and used**. This of course implies that both the **installer and** the **user** are **responsible for water fittings installed, and for any liability** under the Regulations **for any contravention**. More of this later.

What type of installation does NOT need to comply with the regulations?

Well! We have mentioned one type already, 'private water supplies'. For others we need to look at Regulation 2.-(2) and Regulation 2.-(3).

Regulation 2.-(2) is quite explicit about **temporary water supplies.** However, there are conditions to be met for a supply to be exempt, and the exemption applies to **temporary supplies only.**

The Regulations **do not apply to a water fitting installed or used** '– in connection with water supplied for **other than domestic** or **food production purposes,** provided that –'

(a) **the water is metered;**

(b) **the maximum period of use is one month,** or **three months, with written consent** from the water undertaker, and

(c) **backflow prevention devices are in place** to prevent any water from returning 'through the meter.

The term '**other than domestic or food production purposes**' highlighted above is important to this Regulation. If **temporary fittings are to be installed or used** in any way **for domestic purposes or for the production of food**, or is connected to any pipe or fitting used for those purposes, then **the installation WILL need to conform to the Regulations**.

Regulation 2.-(3) quite clearly states that any water fitting or installation that is not connected to the water undertakers main is not required to comply with these regulations. There is **no requirement at present to apply the Water Regulations to installations supplied from a private source**. e.g. from a private well, however, it would be good practice to apply the same standards of installation to ALL water installation work.

It should be remembered however, and Regulation 2.-(3) makes this point, that should any cross connection be made between a supply from a water undertaker, and a supply from any other source, then the regulations will apply to the alternative source. (See also Prevention of cross connection to unwholesome water, Module 7.)

Regulation 2.-(4) exempts fittings that have been installed before the Water Regulations came into force. and it says:

> '*Nothing in these regulations shall require any person to remove, replace, alter, disconnect or cease to use any water fitting that was lawfully installed or used, or capable of being used, before 1st July 1999*'.

An important point here is that this Regulation applies to fittings **lawfully** installed, or used, or capable of being used before 1 July 1999, unless of course there is any waste of water or contamination taking place.

Where there is actual wastage or contamination, the regulations will apply no matter what age the installation.

Now some work for you.

Self-assessment questions (1)

1. Regulation 1.-(1) begins by simply giving the date of commencement of the Regulations. Give the date that the Water Supply (Water Fittings) Regulations 1999 came into force.

   _____

2. Regulation 1.-(2) gives a number of definitions that support the regulations and which includes 'material change of use'.

   Describe 'material change of use' and give one example of its application.

   Material change of use means _____

   _____

   _____

   For example _____

   _____

3. State the FOUR types of premises that are listed under 'material change of use' and which need to be considered when installing water fittings.

   i) _____ *ii)* _____

   iii) _____

   iv) _____

4. Regulation 2.-(1) gives details of certain types of system/fitting that the Regulations apply to. Complete the following statements.

   The Water Regulations apply to:

   ALL _____ _____ *installed or used after 1 July 1999.*

   in premises that are _____ *by the water undertaker.*

   new works and _____ *to existing installations;*

   all permanent _____ *and to some* _____ *installations.*

5. Regulation 2.-(1) states that the Regulations '– apply to any water fitting installed or used –'.

   Bearing this in mind, underline the statement below which best indicates who is responsible for water fittings installed, liability for any contravention of the Regulation.

   i) the installer   *ii) the user*   *iii) both installer and user*

6. Indicate with **Yes** or **No** whether the following statements are correct.

   i) *The Water Regulations* **do not** *apply to installations using water from private supplies.*   Yes ☐ No ☐

   ii) *The Water Regulations* **do** *apply to installations using water from private supplies.*   Yes ☐ No ☐

7. The Regulations do not apply to a temporary water fitting installed or used for water supplied for non-domestic purposes, provided certain conditions are met. List the conditions that should be met:

   Temporary water fittings may be exempt from Water Regulations providing:

   (a) _____

   (b) _____ *or*

   _____ *and*

   (c) _____

   _____

8. Indicate by **Yes** or **No** if the following statements are correct.

   i) *Temporary fittings installed for domestic purposes or food production must comply with the Regulations.*   Yes ☐ No ☐

   ii) *The Regulations do not apply to fittings installed before 1 July 1999.*   Yes ☐ No ☐

   iii) *However, where there is waste or contamination of water the Regulations will apply no matter what the age of the installation.*   Yes ☐ No ☐

   iv) *The Regulations do not apply to installations that are not supplied by the water undertaker unless there is a cross connection with another supply that is supplied by the water undertaker.*   Yes ☐ No ☐

Check your answers on pages 36 and 37.

Part II – Requirements

Part II looks at requirements for the installer and the user (Regulation 3). It looks at requirements for water fittings (Regulation 4). It specifies installations that need to be notified to the water undertaker (Regulation 5) and it sets out basic requirements for approved installer schemes (Regulation 6).

What requirements affect the installer and the user?

Regulation 3

Restriction on installation etc of water fittings.

3.-(1) No person shall:

(a) *install a water fitting to convey or receive water supplied by a water undertaker, or alter, disconnect or use such a fitting; or*

(b) *cause or permit such a water fitting to be installed, altered, disconnected or used, in contravention of the following provisions of this Part.*

(2) No water fitting shall be installed, connected, arranged, or used in such a manner that it causes or is likely to cause:

(i) *waste, misuse, undue consumption or contamination of water supplied by a water undertaker; or*

(ii) *the erroneous measurement of water supplied by a water undertaker.*

(3) No water fitting shall be installed, connected, arranged or used which by reason of being damaged, worn or otherwise faulty, causes or is likely to cause:

(i) *waste, misuse, undue consumption or contamination of water supplied by a water undertaker; or*

(ii) *the erroneous measurement of water supplied by a water undertaker.*

Regulation 3.-(1) requires that **any water fitting conveying or receiving water from a water undertaker's main shall comply with the Requirements of this Part of the Regulations**.

'This Part' means 'Part II – Requirements' and **for this regulation to be satisfied fittings will need to conform to recognised standards**. For example if a fitting complies with a relevant British or European Standard **and** is **properly installed** so that it will not cause any waste, misuse undue consumption, or contamination, or cause or lead to erroneous measurement, it should be accepted as complying with the Regulations.

This point is made more clearly in Regulation 4.

Regulation 3.-(2) and (3) requires water fittings to be in good condition when fitted, and remain in good condition in use. Damaged, worn or otherwise faulty water fittings are not permitted, and the **installation must be installed in a proper manner** so as **not to cause waste, misuse, undue consumption or contamination, or the erroneous measurement of water supplied by the water undertaker.**

Similarly the consumer has a responsibility under Regulation 3 to use water fittings in a proper manner so as **not to cause waste, misuse, undue consumption or contamination, or the erroneous measurement of water supplied by the water undertaker.**

Regulation 3 mentioned a number of terms, some of which seem to have similar meanings.

What is meant by the terms; waste, misuse, undue consumption, contamination and erroneous measurement of water supplied by the water undertaker?

Waste. Water escaping from defective pipes or fittings.

e.g. – dripping tap or leaking pipe

Misuse. Water used for purposes other than that for which it is legally supplied.

e.g. – Fire hose used for washing or cleaning purposes

e.g. – Use of water for filling swimming pool of more than 10,000 litres capacity without having given notification to water undertaker

Undue consumption. Water used in excess of that needed for a specific purpose.

e.g. – automatic flushing cistern set at too high a flow rate

    – WC cistern with larger capacity than that required.

Contamination. The possible contamination of water by unlawful practice.

e.g. – risk of backflow by incorrectly installed pipes/fittings

Erroneous measurement. Wilfully, incorrectly or inaccurately recording a meter reading.

e.g. – Wrongly recorded water meter consumption, due to:

    – bypassing water meter, or

    – interfering with water pipe arrangement so as to affect a meter reading

The above descriptions mention another term, 'water for domestic purposes', which in relation to water supply means water used for drinking, washing, cooking, central heating and sanitary purposes, and in a house or a building used mainly as a house, it also includes the washing of vehicles and the watering of gardens providing the water is drawn off from within the house and without the use of a hosepipe.

(2) *This regulation does not apply to the installation by an approved contractor of a water fitting falling within Paragraph 2, 4(b) or 4(g) in the table.*

(3) *The notice required by Paragraph (1) shall include or be accompanied by:*

    (a) the name and address of the person giving notice, and (if different) the name and address of the person on whom notice may be served under Paragraph (4) below;

    (b) a description of the proposed work or material change of use, and

    (c) particulars of the location of the premises to which the proposal relates, and the use or intended use of those premises;

    (d) except in the case of a fitting falling within Paragraph 4(a), (c), (h) or 5 in the Table above,

        (i) a plan of those parts of the premises to which the proposal relates; and

        (ii) a diagram showing the pipework and fittings to be installed; and

    (e) where the work is to be carried out by an approved contractor, the name of the contractor.

(4) *The water undertaker may withhold consent under Paragraph (1), or grant it subject to conditions, by a notice served before the expiry of the period of 10 working days commencing with the day on which the notice under that paragraph was given.*

(5) *If no notice is given by the water undertaker within the period mentioned in Paragraph (4), the consent required under Paragraph (1) shall be deemed to have been granted unconditionally.*

To whom should notice be given and how?

Regulation 5.-(1) gives details of a number of water fittings installations that the installer must give notice of and sets out a table of operations affected by this regulation. We'll look at the table shortly but before that let's look at a couple of important points relating to the notice that must be given.

– **Notice must be given in writing to the water undertaker,**
– **by the installer** (or his representative) **before commencement of the work,** and
– the **work should not start until consent has been given**, or
– **if no reply has been received after 10 days, consent may be deemed to be given.**

On receipt of the notice, the **water undertaker has 10 days** in which to take one of the following three options:
– to **give consent** to the proposal;
– to **refuse consent; or**
– to **give consent subject to conditions** which the undertaker may see fit to impose.

Whichever action **the water undertaker** chooses to take, he **must be seen to be reasonable** in his decision. Any refusal must be reasonable and any conditions imposed must be reasonable.

It is a requirement of Regulation 5.-(1)(c) that any **conditions set by the water undertaker must be complied with.**

For what type of installation should I give notice?

Notice to the water undertaker **is required where a water fitting is to be installed in connection with** one of the following operations:

1. **the erection of a building or other structure,** not being a pond or swimming pool.

   If a building is being erected and it involves the installation of water fittings notice must be given.

   If the building being erected is to have **no** plumbing installed but needs water for building purposes, then notification will be needed for the temporary building supply, bearing in mind what was said about temporary supplies in Regulation 2.-(2).

2. **the extension or alteration of water fittings on any premises other than a house.** You must notify the water undertaker of any alterations you do to water fittings installations **except** in a house. So! There is no need to give notification for the majority of small repair jobs or replacement work in existing dwellings, but this is of course, subject to sub-paragraph (4) below.

   It should be remembered that 'house' in these Regulations will include a bungalow, flat, maisonette or any premises occupied by a single family as a domestic dwelling.

3. **a material change of use of any premises.**

   Here notification may on occasion be required even without any installation or alteration of water fittings. Remember what was said about 'material change of use' in Regulation 1.-(2). If the use of a building is changed from one category to another, or it is planned to store or use substances that fall into fluid category 4 or 5, then notification is needed.

4. **the installation of any of the following:**

   (a) **a bath** having a capacity, as measured to the centre line of overflow, **of more than 230 litres.**

   (b) **a bidet with an ascending spray or flexible hose.** Over rim types with proper air gaps between tap outlet and overflowing level **do not** need to be notified.

   (c) **a single shower unit** (which may consist of one or more shower heads within a single unit), connected directly or indirectly to a supply pipe which is **of a type specified by the regulator.** Drencher showers installed for reasons of safety or health are exempted from notification. NB: Following a consultation on a definition based on showers with more than one shower head, Ministers have decided not to adopt a definition at this time. This requirement has therefore effectively not be enacted.

   (d) **a pump or booster drawing more than 12 litres per minute,** connected directly or indirectly to a supply pipe.

   (e) **a unit which incorporates reverse osmosis.** This will include home dialysis machines which will require backflow protection.

   (f) **a water treatment unit which produces a waste water discharge or which requires the use of water for regeneration or cleaning.** Base exchange water softeners will be included here.

   (g) **a reduced pressure zone valve assembly** or other mechanical device for protection against a fluid which is in fluid category 4 or 5. (See also Module 6/7)

   (h) **a garden watering system unless designed to be operated by hand.** Operated by hand means a hosepipe actually held in the hand while it is in use.

   (i) **any water system laid outside a building and either less than 750mm or more than 1350mm below ground level.** This is an interesting one! A supply pipe correctly laid from the undertaker's main to a building with no branch pipes rising above ground will be excused notification.

5. **the construction of a pond or swimming pool with a capacity greater than 10,000 litres** which is **designed to be replenished by automatic means** and is to be filled with water supplied by a water undertaker. So! This means large capacity pools but only those that are to be automatically filled. Pools filled by hand are exempted from notification. Incidentally, Paragraph 31 of Schedule 2 requires ponds, fountains and pools to have impervious linings to prevent leakage of water.

Regulation 5.-(2) permits certain installations by an approved contractor to be excused from notification to the water undertaker. Those **installations for which the approved contractor need not give notification**, are:

- **the extension or alteration of a water system on any premises other than a house**

- **a bidet with an ascending spray or flexible hose**, and

- **a reduced pressure zone valve assembly.**

Regulation 5.-(3) gives details of the information that must to be given in the notice to the water undertaker. A wide range of information is expected to be given.

The notice to the undertaker **should contain full details** of the installer or approved contractor, a description **of the work** and its location, the use of the building **including** in some cases, **a plan of the premises and diagrams of the pipework** to be installed.

An example of a suitable 'notice of intention to install water fittings' to water undertaker is shown on the next page. It is likely that individual water undertakers will produce such forms for you to use, but in the absence of one from them, a copy of the form shown overleaf could be used.

It would always be advisable to keep a duplicate copy of any forms sent to the water undertaker to help avoid any misunderstanding at a later date.

NOTICE OF INTENTION TO INSTALL WATER FITTINGS

I hereby give notice as required under Regulation 5 of the Water Supply (Water Fittings) Regulations 1999 that I intend to install water fittings as follows:

Intended installation date ☐

Location of premises where work is to be done _____

_____

Use of building to which the notice refers _____

_____

Description of proposed work/fittings _____

_____

_____

Is plan of proposed installation included?   Yes ☐   No ☐

Will there be a material change of use of the premises?   Yes ☐   No ☐   If Yes give details

_____

_____

Name of installer _____ Approved Contractor Number _____

Company name and address _____

_____

Name of person on whom notice may be served (if different to above) _____

and address _____

_____

Signed _____ Date _____

Regulation 5.-(4) The water undertaker has **10** working **days in which to give consent**, or if he feels the proposed installation does not comply with the regulations **withhold his consent**, or he may **give consent subject to conditions**. e.g. He may insist on a meter for a high consumption shower installation.

The water undertaker must give written notice of his approval or disapproval within 10 working days.

If no reply has been received from the water undertaker **by the end of 10 days**, it can be considered that **consent is granted without conditions.** (Regulation 5.-(5))

The implication here is that notice should be given by the installer at least 10 days before intending to commence work so as to give the water undertaker enough time to consider and perhaps act on the notification within the 10 days he is given to reply, otherwise the start date of installation could be delayed.

The Regulations mention a Contractor's certificate
What is that all about?

Regulation 6

Contractor's certificate

6.-(1) *Where a water fitting is installed, altered, connected or disconnected by an approved contractor, the contractor shall upon completion of the work furnish a signed certificate stating whether the water fitting complies with the requirements of these Regulations to the person who commissioned the work.*

(2) *In the case of a fitting for which notice is required under Regulation 5(1) above, the contractor shall send a copy of the certificate to the water undertaker.*

This Regulation is about self certification. It encourages the wide spread introduction of a **voluntary scheme of approval** for plumbing installers (contractors). The idea is not new, as some water undertakers were operating schemes under water byelaws prior to the introduction of the new Regulations. The scheme may be run by individual water undertakers, but based on a national guidance framework. Other organisations can also submit proposals to be allowed to approve people for the scheme.

The scheme will enable approved contractors to carry out their own inspections and **self-certify their installation work** with out having to involve the water undertakers inspectors. However the undertaker will need to monitor the scheme and audit a sample of contractor's work to ensure that the installer is continuing to comply with the Regulations.

In order to gain approval under the scheme an installer will need to show evidence of competence, i.e. knowledge, experience, training and professional qualifications. In some cases the installer may need to be trained in knowledge of the regulations and others may be required to sit an entrance test.

A list of approved contractors will be maintained by the Water Undertaker who will provide written approval to each individual for each individual accepted by them onto the scheme, and for the type/s of work for which they were accepted.

An approved contractor is required under the scheme to complete a certificate of compliance stating that the work conforms to the requirements of the Water Regulations, and which must be forwarded:

– to the person for whom the work is done, and
– where notice is required under Regulation 5.-(1) a copy must be sent **to the water undertaker.**

Self-assessment questions (2)

1. The basic requirements for the installation of water fittings are given in the following statements which are incomplete. Please fill in the missing words:

   a) Regulation 3.-(1) says in effect, that any water fitting conveying or receiving water from the undertaker's main shall comply with the_____ of Part II of the Regulations'.

   b) Regulation 3.-(2) requires water fittings to be in good condition when fitted and installed in a proper manner so as not to cause

   i) _____, _____, undue consumption or

   _____, of water supplied by the water undertaker or

   ii)   the _____, _____, of water supplied by the water undertaker.

2. Complete the following statements.

   Regulation 4.-(1) states quite clearly that every water fitting shall:

   (a) be of an appropriate _____ and standard;

   (b) be suitable for the _____ in which it is used.

3. There are FOUR ways [given in Reg,4.(2)] in which a fitting can be shown to be of an appropriate quality and standard, state what they are.

   a) _____

   b) _____

   c) _____

   d) _____

4. Similarly, Regulation 4.-(4) requires water fittings to be installed in a 'workmanlike' manner. Explain briefly what this means.

   _____

   _____

   _____

   _____

5. Regulation 4.-(2) and 4.-(5) refers to a 'specification approved by the regulator.' State who is the regulator (i) in England and (ii) in Wales.

  i)   in England_____

  ii) in Wales_____

6. Regulation 5.-(1) is about the requirement to notify water undertakers of certain water fittings installations. State:

  a) **who should give notice**

  _____

  b) **when the notice should be given**

  _____

  c) **when work may proceed**

  _____

7. Answer Yes or No to the following statements regarding notification to water undertakers

  a) i)   Notice is required to be given in writing   Yes ☐   No ☐

     ii) Notice may be given verbally   Yes ☐   No ☐

  b) Notice is required for:

     i)   the erection of a building or other structure;   Yes ☐   No ☐

     ii) the extension or alteration of water fittings on any premises
          other than a house: Yes ☐   No ☐

     iii) a material change of use of any premises;   Yes ☐   No ☐

  c) Where work is **not** being carried out by an approved contractor, notice is required for the installation of:

     i)   a bath of more than 230 litres capacity   Yes ☐   No ☐

     ii) an ascending spray type bidet   Yes ☐   No ☐

     iii) an ascending spray type bidet   Yes ☐   No ☐

d) Notice is required for:

    i)   a single shower unit   Yes ☐   No ☐

    ii)   a drencher shower for safety or health purposes   Yes ☐   No ☐

e) Notice is required for:

    i)   every pump drawing water from the water from the undertaker's main   Yes ☐   No ☐

    ii)   a pump drawing more than 12 litres per minute of water from the water undertaker's main   Yes ☐   No ☐

    iii) a home dialysis machine   Yes ☐   No ☐

    iv) a base exchange water softener   Yes ☐   No ☐

    v)   machinery for washing vehicles   Yes ☐   No ☐

    vi) a reduced pressure zone (RPZ) valve assembly   Yes ☐   No ☐

f) Notice is required for a garden watering system:

    i)   designed to be operated by hand   Yes ☐   No ☐

    ii)   designed for automatic operation   Yes ☐   No ☐

g) Notice is required for an outside water system:

    i)   laid at a depth of 700mm below ground   Yes ☐   No ☐

    ii)   laid at a depth of 800mm below ground   Yes ☐   No ☐

    iii) laid at a depth of 1400mm below ground   Yes ☐   No ☐

    iv) laid above ground level   Yes ☐   No ☐

    v)   the construction of a pond or swimming pool with a capacity of not less than 10,000 litres which is to be filled with water supplied by a water undertaker;   Yes ☐   No ☐

8. Regulation 5.-(2) says that 'approved contractors' do not need to give prior notification to the water undertaker for THREE types of installation.

State what the THREE installations are:

1. _____

2. _____

3. _____

9. Water undertakers are to be encouraged under Regulation 6 to introduce an approved contractors scheme within their area of supply.

Show by Yes or No whether the following statements are correct.

a) *Contractors are required to join the scheme*   Yes ☐   No ☐

b) *Contractors may join the scheme voluntarily*   Yes ☐   No ☐

c) *The scheme will introduce a degree of self certification*   Yes ☐   No ☐

10. An approved contractor is required to complete a certificate of compliance.

a) *What should be stated on the certificate of compliance?*

_____

b) *State who the certificate should be given to when completed.*

_____

c) *Under what circumstances should a copy of the certificate of compliance be forwarded to the water undertaker?*

_____

Check your answers on pages 38, 39 and 40.

Part III – Enforcement etc

Regulation 7

Penalty for contravening regulations

7.-(1) Subject to the following provisions of this regulation, a person who:

(a) contravenes any of the provisions of regulations 3(1), (2) or (3) or 6(1) or (2);

(b) commences an operation listed in the table in regulation 5(1) without giving notice required by that paragraph;

(c) commences an operation listed in the table in regulation 5(1) without the consent required by that paragraph; or

(d) carries out an operation listed in the table in regulation 5(1) in breach of a condition imposed under regulation 5(4);

is guilty of an offence and liable on summary conviction to a fine not exceeding level 3 on the standard scale.

(2) In any proceedings against an owner or occupier for an offence under paragraph

(1) which is based on installation, alteration, repair, connection or disconnection of a water fitting, it shall be a defence to prove:

(a) that the work in question was carried out by or under the direction of an approved contractor; and

(b) that the contractor certified to the person who commissioned the work that the water fitting complied with the requirements of these Regulations.

Regulation 7.-(1) is quite straightforward in that it sets out the amount of fine that is applicable on conviction for an offence under these regulations.

It says in effect that **anyone who installs or uses water fittings, to cause waste, misuse, undue consumption, or contamination of water, will be contravening the regulations**. Likewise **anyone who fails to give notice** of intended work, or **commences the work without consent** from the water undertaker, or **fails to comply with any conditions** set by the undertaker **is also contravening the regulations and will be liable to conviction**.

- **A person who contravenes the regulations,**
- **may, on summary conviction,**
- **be subjected to a fine not exceeding level 3 on the standard scale.**

It will be unusual for just one offence to be committed. Usually by breaking one regulation, others will automatically be broken. So fines could be in multiples of the level 3 fine.

Note: In Scotland the potential fine is not exceeding level 5.

Regulation 7.-(2) Remember we said earlier that both the installer and the user were responsible for, and may be liable for offences committed under the Regulations. This regulation provides for owners or occupiers of premises to be excused from their responsibility, but there are conditions attached.

An owner or occupier of premises may be excused responsibility, if:

– the **work** in question **was installed**, altered, connected or disconnected **by, or under the direction of, a contractor approved by the water undertaker, and the contractor certified** (to the person who commissioned the work) **that the water fitting was in compliance with the Regulations**.

Regulation 8

Modification of Section 73 of the Act

8. In Section 73 of the Act (offences of contaminating, wasting and misusing water etc), after sub-section (1) there shall be inserted:

"*(1A) In any proceedings under subsection (1) above it shall be a defence to prove:*

(a) *That the contamination or likely contamination, or the wastage, misuse or undue consumption, was caused (wholly or mainly) by the installation, alteration, repair or connection of the water fitting on or after 1st July 1999;*

(b) *that the works were carried out by or under the direction of an approved contractor within the meaning of the Water Supply (Water Fittings) Regulations 1999; and*

(c) *that the contractor certified to the person who commissioned those works that the water fitting complied with the requirements of those Regulations.*"

This regulation is included for legal purposes.

Before Regulation 7 could be implemented, it was necessary to make provision for it by first amending Section 73 of the Water Industry Act. This now takes into account the Approved Contractor and the Approved Contractor's Certification requirements.

Regulation 9

9. *Inspections measurements and tests*

Any person designated in writing:

(a) for the purposes of Section 74(4) or 170(3), by a water undertaker, or

(b) for the purposes of Section 84(2), by any local authority,

may carry out such inspections, measurements and tests on premises entered by that person or on water fittings or other articles found on any such premises, and take away such samples of water or of any land, and such water fittings and other articles, as that person may consider necessary for the purposes for which those premises were entered.

We are all aware of the water inspector's role in checking that our installations comply with the Regulations but this regulation takes the role a step further.

Certain representatives from water undertakers and local authorities (Town or District Councils) **who are authorised in writing, may enter premises to carry out inspections, measurements and tests, and are permitted to take away samples** for the purposes of the Regulations.

Samples may be taken if need be to confirm that a material complies with the Regulations or otherwise as the case may be, For example, could mean a piece of pipe or a fitting that is thought not to be to the required quality or standard, or it could be a sample of water taken from the system that is thought to be contaminated.

The term 'any person designated in writing' means a representative of either the water undertaker or a local authority who has been authorised by them to enter premises and can show proof of identity from the water undertaker or the local authority.

Regulation 10

Enforcement

10.-*(1)* *A water undertaker shall enforce the requirements of these regulations in relation to the area for which it holds an appointment under Part II of the Act.*

(2) *The duty of a water undertaker under this regulation shall be enforceable under Section 18 of the Act:*

(a) by the regulator; or

(b) with the consent of or in accordance with a general authorisation given by the regulator, the Director.

In Regulation 10.-(1) **the responsibility for enforcement** of the regulations **is placed** quite firmly **on the water undertaker**, an obligation which it may carry out either by using inspectors employed by the undertaker, or plumbers who are approved contractors under an approved contractors scheme.

In Regulation 10.-(2) you will see that **it is a duty of the regulator to ensure that water undertakers** carry out their duty to enforce the Regulations.

Also in Regulation 10.-(2) **the Director has a duty to see that water undertakers enforce the Regulations**.

'The Director' is: **The Director General of Water Services** (OFWAT).

Regulation 11

Relaxation of requirements

11.-(1) *Where a water undertaker considers that any requirement of Schedule 2 to these regulations would be inappropriate in relation to a particular case the undertaker may apply to the regulator to authorise a relaxation of that requirement.*

(2) *The water undertaker shall give notice of any proposed relaxation in such manner and to such persons as the regulator may direct.*

(3) *The regulator may grant the authorisation applied for with such modifications and subject to such conditions as he thinks fit.*

(4) *The regulator shall not grant an authorisation before the expiration of one month from the giving of the notice, and shall take into consideration any objection which may have been received by him.*

(5) *A water undertaker to whom an authorisation is granted under Paragraph (3) in a particular case may relax the requirement of Schedule 2 in that case in accordance with the terms of that authorisation.*

There are occasions when, or situations in which, the regulations are difficult to apply. In these cases, under Regulation 11, **the water undertaker has the option, if it thinks it necessary, to apply to the regulator for a relaxation of the Regulations**.

If the regulator thinks their case is a reasonable one, (after consultation) he will permit the water undertaker to apply different rules or to vary the rules to suit the particular situation.

An example of this may be where a so called 'street loo' containing WC mechanism of the drop-valve type is installed prior to 1 January 2001. The street loo is designed to give access to the disabled and is therefore a special case. The provision of drop-valve mechanisms are permitted in post 2001 installations. Prior to 2001 a relaxation would be required for this type of appliance to be installed.

Relaxations will not be easy to get and there will be no 'blanket' relaxation. Any application will be treated according to its merits and one relaxation will not automatically lead the way to another similar application.

Regulation 12

Approval by regulator or water undertaker

12.-(1) *Before approving a specification under Regulation 4 or under Schedule 2, the regulator shall consult:*

    (a) *every water undertaker*

    (b) *such trade associations as he considers appropriate, and*

    (c) *such organisations appearing to him to be concerned with the interests of water users as he considers appropriate.*

(2) *Where the regulator approves a specification under regulation 4 or under Schedule 2, he shall give notice of the approval to all persons who were consulted under Paragraph (1) and shall publish it in such a manner as he considers appropriate.*

(3) *Where the water undertaker approves a method of installation under regulation 4, the water undertaker shall give notice of the approval to the regulator and shall publish it in such a manner as the undertaker considers appropriate.*

(4) *This regulation applies to the revocation or modification of an approval as it applies to the giving of that approval.*

The overall responsibility for making and enforcing the Water Regulations lies with the regulator, and he does this by using experts appointed to the Water Regulations Advisory Committee to advise civil servants, who in turn, carry out the work on his behalf.

In England the regulator is of course the Secretary of State for the Department of the Environment, Food and Rural Affairs, and in the case of Wales, the National Assembly of Wales.

Regulation 12 places an obligation on the regulator to consult with various organisations representing water users before giving the Regulations his approval, and before the Regulations are published and enforced.

This obligation also applies to any amendments that might be made to the Regulations and to any other approved documents relating to the Regulations.

The regulator will consult with:
– the water industry,
– the plumbing and mechanical services industry, and
– others who use water and who have an interest in how water is used.

Under Regulation 12(3) and (4), if a water undertaker approves a method of installation (as it is permitted to do) it is required to notify the regulator of its decision and to publish it in some suitable way to show that it has allowed that method of installation to be used.

By notifying the regulator and publishing its actions, the water undertaker can be seen to be fair and reasonable in its actions, and of course will be open to challenge if he is not!

Regulation 13

Disputes

13. Any dispute between a water undertaker and a person who has installed or proposes to install a water fitting:

 (a) as to whether the water undertaker has unreasonably withheld consent, or attached unreasonable conditions, under Regulation 5 above: or

 (b) as to whether the water undertaker has unreasonably refused to apply to the regulator for a relaxation of the requirements of these regulations,

 shall be referred to arbitration by a single arbitrator to be appointed by agreement between the parties or, in default of agreement, by the regulator.

It is inevitable that occasionally there will be a disagreement between water undertaker and installer over the application and enforcement of the regulations.

In most cases these disagreements can be settled quite amicably between them.

During any discussions relating to a disagreement there are two important questions to be asked.

1. **Does the installation fully comply with the Regulations** and its requirements?

2. **Is the Water Undertaker being reasonable** in his refusal to accept the installation or proposed installation?

If both parties keep these questions in mind disputes will not last long and agreement should be relatively easy to reach.

However, there will be occasions when agreement cannot be reached between the two parties. **Regulation 13** has been written to cover such cases which may be referred to arbitration by an arbitrator appointed by the regulator. The arbitrator will be asking the same two questions as are given above.

Should there still be disagreement after the arbitrator has tried to resolve the dispute, the case will go to the regulator for a final decision.

Regulation 14

Revocation of byelaws

14. *The byelaws referred to in column (2) of Schedule 3, being made or having effect as if made by the water undertakers referred to in column (1) of Schedule 3 under Section 17 of the Water Act 1945(a), are hereby revoked.*

As you can see this regulation simply refers to Schedule 3 which lists all the water undertakings affected by these regulations and the dates that their previous byelaws came into effect.

All of these are now replaced by the Water Supply (Water Fittings) Regulations 1999. Schedule 3 is copied below for your information.

SCHEDULE 3

Regulation 14

BYELAWS REVOKED

Water undertaker	Byelaws
(1)	(2)
Anglian Water Services Ltd.	The byelaws made on 1st October 1987
Bournemouth and West Hampshire Water plc.	The byelaws made on 14th January 1987
Bristol Water plc.	The byelaws made on 23rd February 1987
Cambridge water plc.	The byelaws made on 29th January 1987
Dee Valley water plc.	(a) the byelaws made on 29th January 1987 by the Chester Water Company
	(b) the byelaws made on 3rd January 1988 by the Wrexham Water Company
Dwr Cymru Cyngedig.	The byelaws made on 6th December 1988
Essex & Suffolk Water plc.	The byelaws made on 1st May 1987
Folkstone & Dover Water Services plc.	The byelaws made on 27th March 1987
Hartlepool Water plc.	The byelaws made on 20th March 1987
Mid Kent Water plc.	The byelaws made on 24th March 1987
North Surrey Water Ltd.	The byelaws made on 26th February 1987
North West Water Ltd.	The byelaws made on 12th March 1987
Northumbrian Water Ltd.	The byelaws made on 7th April 1987
Portsmouth Water plc.	The byelaws made on 26th February 1987
Severn Trent Water Ltd.	The byelaws made on 13th March 1987
South Staffordshire Water plc.	The byelaws made on 18th March 1987
South East Water Ltd.	The byelaws made on 9th March 1987
South West Water Services Ltd.	The byelaws made on 8th April 1987
Southern Water Services Ltd.	The byelaws made on 2nd April 1987
Sutton and East Surrey Water plc.	The byelaws made on 13th March 1987
Tendring Hundred Water Services Ltd.	The byelaws made on 31st March 1987
Thames Water Utilities Ltd.	The byelaws made on 3rd April 1987
Three Valleys Water Services plc.	The byelaws made on 24th March 1987
Wessex Water Services Ltd.	The byelaws made on 16th January 1987
York Waterworks plc.	The byelaws made on 11th March 1987
Yorkshire Water Services Ltd.	The byelaws made on 17th March 1987

Self-assessment questions (3)

1. Regulation 7.-(1) sets out the penalties which may be applied for contravention of the Regulations. Complete the following statements:

   A person who contravenes the regulations may,
   on conviction, be subjected to a fine not exceeding _____ *on the standard scale.*

2. Regulation 7.-(2) provides owners or occupiers of premises with a defence, should there be a prosecution against them for contravention of the Regulations, but there are conditions attached. Complete the statement below to give the conditions:

   An owner or occupier may have a defence against prosecution if:

   _____

   _____

   _____

3. Complete the following statement:

   Representatives from water undertakers and _____ *who*
   are authorised to enter premises to carry out inspections, measurements and tests are also permitted to
   _____ *for the purposes of the Regulations.*

4. State:

   a) *who has the responsibility for enforcement of the regulations*

   _____

   b) *who is responsible for ensuring that enforcement is carried out*

   _____

5. Regulation 13 deals with disputes that might arise between undertakers and installers.

   In the event that a dispute cannot be resolved locally, give:
   a) *the person who has been appointed by the regulator to settle the dispute, and*
   b) *the TWO basic principles he will consider when making his decision.*

   a) *The person is* _____

   b) i) _____

   ii) _____

6. Answer Yes or No depending if you think the following sentence is correct.
   Regulation 14 revokes any water byelaws made under Section 17 of the Water Act 1945.   Yes ☐   No ☐

Check your answers on page 41.

Summary of main points

Part I

The Regulations came into force on 1 July 1999. [Regulation 1.-(1)]

The Water Regulations apply to: [(Regulation 2.-(1)]

- **the Regulations apply to all water fittings installed or used** after 1 July 1999;
- **in ALL premises that are supplied by the water undertaker.**
- **new works, and alterations to existing work,** including **connections and disconnections;**
- **all permanent installations** and to **some temporary installations.**

Both the **installer and** the **user** are **responsible for water fittings installed, and for any liability** under the regulations **for any contravention.**

There is no requirement at present to apply the regulations to installations supplied from a private source. However, it would be good practice to apply the same standards of installation to ALL water installation work.

Regulation 2.-(2) The Regulations **do not apply to temporary water fittings installed** or used for water supplied **for other than domestic purposes, provided that:**

(a) **the water is metered;**

(b) **the maximum period of use is one month, or three months, with written consent** from the undertaker, **and**

(c) **backflow prevention devices are in place** to prevent any water from returning back through the meter.

However, temporary fittings installed to be used in any way for **domestic purposes, WILL need to comply with the Regulations.**

Regulation 2.-(3) **exempts fittings** that have been **lawfully installed before the Water Regulations came into force** (1 July 1999), unless of course there is any **actual waste, misuse, undue consumption or contamination, or the erroneous measurement of water supplied** taking place. (Regulation 2.-(4))

Part II

Any water fitting conveying or receiving water from the undertaker's main **shall comply with the Requirements of Part II of these Regulations**. [Regulation 3.-(1)]

For the Regulations to be satisfied, **fittings will need to conform to recognised British or European standards (**e.g. BS**) and be properly installed**.

Regulation 3.-(2) **requires water fittings to be in good condition when fitted, and remain in good condition in use**. Damaged, worn or otherwise faulty water fittings are not permitted, and the **installation must be installed** so as **not to cause waste, misuse, undue consumption or contamination, or the erroneous measurement of water supplied by the undertaker**.

Every water fitting shall: [Regulation 4.-(1)]

(a) **be of an appropriate quality and standard;**

(b) **be suitable for the circumstances in which it is used.**

Regulations 4.-(2) The fitting must show evidence that:

(a) it **bears an appropriate CE Mark** in accordance with the Directive;

(b) it **conforms to an appropriate harmonised standard or European technical approval;**

(c) **it conforms to an appropriate British Standard or British Board of Agrément Certificate,** or to some other national specification of any EEA State which provides an equivalent level of protection and performance; or

(d) it **conforms to a specification approved by the regulator**.

Regulation 4.-(3) states '**Every water fitting shall comply with the requirements of Schedule 2 to these regulations'.** The requirements set out in the Schedule 2 are, in effect, the rules we must abide by when we install water fittings and must be followed as though they were actually written as regulations.

Regulation 4.-(5) requires **water fittings to be installed in a 'workmanlike manner'.**

Regulation 4.-(6) A water fitting is installed in a workmanlike manner if its installation conforms to:

(a) **to an appropriate British Standard, a European technical approval or some other national specification of an EEA State** which provides an equivalent level of protection and performance, **or**

(b) **to a specification approved by the regulator.**

(c) **to a method of installation approved by the water undertaker**

Approved specifications are those that **have the approval of the regulator**.
- **In England the regulator is the Secretary of State for the DETR**
- **In Wales the regulator is the National Assembly of Wales**

Regulation 5.-(1) requires **that notice shall be given in writing, by the installer, to the water undertaker, before commencement of the work,** and **work should not proceed until consent has been given** or after 10 days if no reply has been received.

Notice is required where a water fitting is to be installed in connection with:

1. **The erection of a building or other structure;**

2. **The extension or alteration of water fittings on any premises other than a house;**

3. **A material change of use of any premises;**

4. **The installation of:**

    (a) **a bath of more than 230 litres capacity;**

    (b) **a bidet with ascending spray or flexible hose;**

    (c) **a single shower unit** (with one or more shower) **of a type approved by the regulator;** (drencher showers installed for reasons of safety are excepted)

    (d) **a pump or booster drawing more than 12 litres per minute;**

    (e) **a unit which incorporates reverse osmosis;**

    (f) **a water treatment unit which produces a waste water discharge or which requires the use of water for regeneration or cleaning;**

    (g) **a reduced pressure zone (RPZ) valve assembly;**

    (h) **a garden watering system unless designed to be operated by hand;**

    (i) **any water system laid outside a building and either less than 750mm or more than 1350mm below ground level;**

5. **The construction of a pond or swimming pool with a capacity greater than 10,000 litres, designed to replenished by automatic means.**

The notice to the undertaker **should contain full details** of the installer or approved contractor, a description **of the work** and its location, the use of the building **including** in some cases, **a plan of the premises and diagrams of the pipework** to be installed.

Approved contractors need not notify the water undertaker of the installation of:

2   the extension or alteration of water fittings on any premises other than a house

4(b)   a bidet with ascending spray or flexible hose; or

4(g)   a reduced pressure zone (RPZ) valve assembly

Regulation 5.-(3) The water undertaker must give approval (or otherwise) **of notifications within 10 working days** of receipt of the notice. **If no reply has been received by the end of 10 days consent is deemed to be granted without any conditions.** (Regulation 5.-(5))

Regulation 6.- Approved contractors are required to complete a certificate of compliance stating that the work conforms to the requirement of the regulations **which must be forwarded to the person for whom the work was done. Where notice is required under Regulation 5(1) a copy of the notice should be sent to the water undertaker.**

Part III

Regulation 7.-(1) sets out the amount of fine that may be applied.

A person who contravenes the regulations may on summary conviction;

be subjected to a fine not exceeding level 3 on the standard scale.

Regulation 7.-(2) **An owner or occupier of premises may be excused liability, but only if the water fitting** in question **was installed, altered, connected or disconnected by,** or under the direction of, **an approved contractor, and the contractor has certified** to the owner **that the water fitting was in compliance with the regulations**.

Designated representatives, from water undertakers and local authorities, are authorised under Regulation 9 **to enter premises to carry out inspections, measurements and tests and to take samples for the purposes of the Regulations.**

In Regulation 10(1) **the responsibility for enforcement is placed on the water undertaker**, and Regulation 10.(2) says that **it is a duty of the regulator** (or the Director of OFWAT) **to ensure that water undertakers** carry out their duty to **enforce the Regulations**.

Under Regulation 11 the water undertaker has the option, if he thinks it necessary, to apply to the regulator for a relaxation of the regulations, but relaxations will not be easy to obtain.

Disputes (Regulation 13) will in most cases be settled quite amicably between the undertaker and the installer.

However, **when agreement cannot be reached** between the water undertaker and the installer, **disputes may be referred to arbitration** and be settled by an **arbitrator** appointed by the Secretary of State (or by the Director of OFWAT).

The arbitrator will be considering two basic questions.

1. **Does the installation fully comply with the regulations** and its requirements?

2. **Is the Water Undertaker being reasonable** in his refusal to accept the installation or proposed installation?

Answers to self-assessment questions (1)

1. Regulation 1.-(1) begins by simply giving the date of commencement of the Regulations.

   Give the date that the Water Supply (Water Fittings) Regulations 1999 came into force.

   They came into force on 1 July 1999.

2. Regulation 1.-(2) gives a number of definitions that support the regulations and which includes 'material change of use'. Describe 'material change of use' and give one example of material change of use.

   Material change of use **means a change in the purpose for which, or the circumstances in which, premises are used.**

   For example – **a house that is turned into a nursing home, or**

        – **a church that is turned into a dwelling.**

3. State the FOUR types of premises that are listed under 'material change of use' and which need to be considered when installing water fittings.

   i) a dwelling   *ii) an institution*   *iii) a public building*

   iv) a building used for storage or use of substances of fluid category 4 or 5

4. Regulation 2.-(1) gives details of certain types of system/fitting that the Regulations apply to. Complete the following statements.

   The Water Regulations apply to:
   – *ALL* **water fittings** *installed or used after 1 July 1999, in premises that are* **supplied by** *the water undertaker;*
   – *new works and* **alterations** *to existing installations;*
   – *all permanent* **installation**s *and to some* **temporary** *installations.*

5. Regulation 2.-(1) states that the Regulations '– apply to any water fitting installed or used –'.

   Bearing this in mind, underline the statement below which best indicates who is responsible for water fittings installed, liability for any contravention of the Regulation.

   i) the installer   *ii) the user*   *iii)* **both installer and user**

6. Indicate whether the following statements are correct.

i) The Water Regulations **do not** apply to installations using water from private supplies Yes ✓ No ☐

ii) The Water Regulations do apply to installations using water from private supplies Yes ☐ No ✓

7. The Regulations do not apply to a temporary water fitting installed or used for water supplied for non-domestic purposes, provided certain conditions are met. List the conditions:

a) **the water is metered;**

b) **the maximum period of use is one month, or three months, with written consent from the undertaker, and**

c) **backflow prevention devices are in place to prevent any water from returning through the meter.**

8. Indicate by Yes or No if the following statements are correct.

i) Temporary fittings installed for domestic purposes, must comply with the Regulations Yes ✓ No ☐

ii) The Regulations do not apply to fittings installed **before** 1 July 1999 Yes ✓ No ☐

iii) However, where there is waste or contamination of water the Regulations will apply no matter what the age of the installation Yes ✓ No ☐

iv) The Regulations do not apply to installations that are not supplied by the water undertaker unless there is a cross connection with another supply that is supplied by the water undertaker Yes ✓ No ☐

Answers to self-assessment questions (2)

1. The basic requirements for the installation of water fittings are given in the following statements which are incomplete. Please fill in the missing words:

   a) *Regulation 3.-(1) says that any water fitting conveying or receiving water from the undertaker's main shall comply with the* **Requirements** *of Part II of Regulations'.*

   b) *Regulation 3.-(2) requires water fittings to be in a good condition when fitted and water supplied by the water undertaker or installed in a proper manner so as not to cause*

      i) **waste, misuse,** *undue consumption,* **contamination** *of*

      ii) *the* **erroneous measurement** *of water supplied by the water undertaker.*

2. Complete the following statements.

   Regulation 4.-(1) states quite clearly that every water fitting shall;

   a) *be of an appropriate* **quality** *and standard;*

   b) *be suitable for the* **circumstances** *in which it is used.*

3. There are **FOUR** ways [given in Regulation, 4.(2)] in which a fitting can be shown to be of an appropriate quality and standard, state what they are.

   The fitting must show evidence that:

   a) **it bears an appropriate CE Mark;**

   b) **it conforms to an appropriate harmonised standard or European technical approval;**

   c) **it conforms to an appropriate British Standard or some other national specification of an EEA State;**

   d) **it conforms to a specification approved by the regulator.**

4. Regulation 4.-(5) requires water fittings to be installed in a 'workmanlike' manner. Explain briefly what this means.

   A water fitting is installed in a workmanlike manner if its installation conforms to:

   a) **an appropriate British or European Standards, or;**

   b) **an appropriate national specification from another country within the European community;**

   c) **a British Board of Agrément Certificate, or;**

   d) **a specification approved by the Secretary of State.**

5. Regulation 4.-(2) and 4.-(5) refers to a 'specification approved by the regulator'. State briefly who is the regulator (i) in England and (ii) in Wales.

   i) *in England the regulator is* **the Secretary of State for the Department of the Environment, Food and Rural Affairs (DEFRA)**

   ii) *in Wales the regulator is* **the National Assembly of Wales**

6. Regulation 5.-(1) is about the requirement to notify water undertakers of certain water fittings installations. State:

    a) who should give notice   **The installer (or his representative)**

    b) when the notice should be given,   **Before commencement of the work**

    c) when work may proceed   **When consent has been given or after 10 days if no reply has been received**

7. Answer Yes or No to the following statements regarding notification to water undertakers

a) i)   Notice is required to be given in writing   Yes ✔   No ☐

    ii)   Notice may be given verbally   Yes ☐   No ✔

b) Notice is required where water fittings are used in connection with:

    i)   the erection of a building or other structure   Yes ✔   No ☐

    ii)   the extension or alteration of water fittings on any premises other than a house; except where work is carried out by an approved contractor   Yes ✔   No ☐

    iii) a material change of use of any premises   Yes ✔   No ☐

c) Where work is not being carried out by an approved contractor, notice is required for the installation of:

    i)   a bath of more than 230 litres capacity   Yes ✔   No ☐

    ii)   an over rim type bidet   Yes ☐   No ✔

    iii) An ascending spray type bidet   Yes ✔   No ☐

d) Notice is required for:

    i)   a single shower unit   Yes ☐   No ✔

    ii)   a drencher shower for safety or health purposes   Yes ☐   No ✔

e) Notice is required for:

    i)   every pump drawing water from the water undertaker's main   Yes ☐   No ✔

    ii)   a pump drawing more than 12 litres per minute of water from the water undertaker's main   Yes ✔   No ☐

    iii) a home dialysis machine   Yes ✔   No ☐

    iv) a base exchange water softener   Yes ✔   No ☐

    v)   machinery for washing vehicles   Yes ✔   No ☐

    vi) a reduced pressure zone (RPZ) valve assembly   Yes ✔   No ☐

f) Notice is required for a garden watering system:

    i)   designed to be operated by hand   Yes ☐   No ✔

    ii)   designed for automatic operation   Yes ✔   No ☐

g) Notice is required for an outside water system:

    i)   laid at a depth of 700mm below ground   Yes ✔   No ☐

    ii)   laid at a depth of 800mm below ground   Yes ☐   No ✔

    iii) laid at a depth of 1400mm below ground   Yes ✔   No ☐

    iv) laid above ground level   Yes ✔   No ☐

    v)   the construction of a pond or swimming pool with a capacity of not less than 10,000 litres which is to be filled with water supplied by a water undertaker;   Yes ✔   No ☐

8. Regulation 5.-(2) says that 'approved contractors' do not need to give prior notification to the water undertaker for THREE types of installation.

   State what the THREE installations are.

   i)   **the extension or alteration of water fittings on any premises other than a house**

   ii)   **a bidet with an ascending spray or flexible hose**

   iii)   **a reduced pressure zone valve**

9. Water undertakers are to be encouraged under Regulation 6 to introduce an approved contractors scheme within their area of supply.

   Show by Yes or No whether the following statements are correct.

   a) *Contractors are required to join the scheme*   Yes ☐   No ✔

   b) *Contractors may join the scheme voluntarily*   Yes ✔   No ☐

   c) *The scheme will introduce a degree of self certification*   Yes ✔   No ☐

10. An approved contractor is required to complete a certificate of compliance.

    a) *What should be stated on the certificate of compliance?*

       That the work complies with the requirements of the regulations

    b) *State who the certificate should be given to when completed.*

       To the person for whom the work has been done

    c) *Under what circumstances should a copy of the certificate of compliance be forwarded to the water undertaker?*

       Where notice is required under Regulation 5(1)

Answers to self-assessment questions (3)

1. Regulation 7.-(1) sets out the penalties which may be applied for contravention of the Regulations. Complete the following statements:

   A person who contravenes the regulations may, on conviction, be subjected to a fine not exceeding **level 3** *on the standard scale.*

2. Regulation 7.-(2) provides owners or occupiers of premises with a defence, should there be a prosecution against them for contravention of the Regulations, but there are conditions attached. Complete the statement below to give the conditions:

   An owner or occupier may have a defence against prosecution if:

   the water fitting was installed, altered, connected or disconnected by, or under the direction of, a contractor approved by the water undertaker, and the contractor certified that the water fitting was in compliance with the Regulations.

3. Complete the following statement:

   Representatives from water undertakers and **Local authorities** *who are also authorised to enter premises to carry out inspections, measurements and tests are permitted to* **take away samples** *for the purposes of the Regulations.*

4. State: a) who has the responsibility for enforcement of the regulations

   The Water undertaker

   b) who is responsible for ensuring that enforcement is carried out

   The regulator, or the Director *(if authorised by the regulator)*

5. Regulation 13 deals with disputes that might arise between undertakers and installers. In the event that a dispute cannot be resolved locally, give:

   a) the person who has been appointed by the regulator to settle the dispute, and

   b) the TWO basic questions he will consider when making his decision.

   a) *The person is* **The Arbitrator**

   b) *The two basic principles he will consider are:*

       i)   **Does the installation fully comply with the regulations?**

       ii)   **Is the Water Undertaker being reasonable?**

6. Answer Yes or No depending if you think the following sentence is correct.

   Regulation 14 revokes any water byelaws made under Section 17 of the Water Act 1945   Yes ✓   No ☐

What to do next

You have completed your second self learning module

It was rather a long one but I hope you will have found it interesting.

The following modules are concerned with the Requirements of the Regulations which are the 'nuts and bolts' of compliance.

They look more closely at how to ensure your installations comply with the Regulations and its Requirements and most of the following modules are shorter.

Please go on to **Module 3** Materials and substances in contact with water.

Water Industry Act 1991:

Water Supply (Water Fittings) Regulations 1999

An Open Learning Course

Module 3

Materials and substances in contact with water

Introduction

The purpose of Paragraph 2 of Schedule 2 is to **ensure that** the materials we use **will not cause any contamination that will affect the drinking qualities of the water**.

Materials used in water systems should not contain any substance that might be absorbed (leached) into the water to cause the water to be toxic, or biologically unhealthy, or to affect its colour, taste, odour. Materials should not, in any way, cause the water supplied by the undertaker to become unfit to drink.

Materials include metallic and non-metallic substances (including plastics) used in the manufacture of pipes, fittings, and appliances and those used in jointing processes and for protective coatings. The list is almost endless, but suffice to say, if a material or substance is in contact with water it must not adversely affect the quality of our water.

Modules 7 and 8 will discuss water quality further.

It might be considered that substances used to produce materials, fittings and appliances etc are more the concern of the manufacturer than the installer. This is true to a large extent, because they are the people who put the substances into the materials.

Manufacturers **do** have a responsibility and most of them take that responsibility seriously.

But! You also have a responsibility!

As the person who uses the materials, you should be in a position to know which materials are permitted to be used and which of a wide variety of substances and materials you should choose for your particular installation.

Besides! As the installer, it is you who could be prosecuted for installing a harmful substance, not the manufacturer.

So! On to the learning exercise!

It is the aim of **this** module to help you decide which materials are likely to cause contamination, and which ones are approved for use, and of course, how you can be sure that you are complying with Paragraph 2 of Schedule 2.

What is the requirement?

> **Schedule 2: Paragraph 2:**
>
> **Materials and substances in contact with water.**
>
> **2.-(1)** *Subject to sub-paragraph (2) below, no material or substance, either alone or in combination with any other material or substance or with the contents of any water fitting of which it forms a part, which causes or is likely to cause contamination of water shall be used in the construction, installation, renewal, repair or replacement of any water fitting which conveys or receives, or may convey or receive, water supplied for domestic or food production purposes.*
>
> **(2)** *This requirement does not apply to a water fitting downstream of a terminal fitting supplying wholesome water where:*
>
> (a) *The use to which the water downstream is put does not require wholesome water; and*
>
> (b) *a suitable arrangement or device to prevent backflow is installed.*

First, let's look at some of the wording

Paragraph 2 is concerned with the supply of water for domestic and food production purposes. **'Water for domestic purposes'** means wholesome **water supplied by a water undertaker for general use including drinking.** (See also Glossary of terms)

Paragraph 2 is also concerned with water fittings containing wholesome water.

For example, water drawn from a drinking water tap 'inside a dwelling' is considered to be 'wholesome', but once you fit a hosepipe to that tap, to connect up a washing machine, or a garden hosepipe, the situation may change and the installation may not comply with the regulations.

Tap with hosepipe attached constitutes a hazard because the material of the hose could be contaminated

tap
backflow prevention device
garden spray
supply pipe
hosepipe

Any water passing through the hosepipe is considered to be 'unfit for human consumption':

i) because the material the hosepipe is made from could cause contamination, and

ii) because of the backflow risks associated with its use.

(We will look more closely at backflow prevention in Module 8.)

Sub-paragraph 2.-(2)(b) states that the previous requirement does **not** apply to water fittings **downstream** of a terminal fitting supplying wholesome water and a number of examples are given in the Guidance document. In the diagram below, water in the supply pipe up to the tap (terminal fitting) contains wholesome water, and the hosepipe and water contained in it is considered to be a contamination risk.

BUT! It is important that no water in the hosepipe (downstream) can find its way back through the tap into the supply pipe (upstream).

For this reason a backflow prevention device is required to be fitted as shown. The type of device will depend on the severity of the risk.

Hosepipes, flushing cisterns, feed and expansion cisterns, closed circuits, and overflow pipes or warning pipes are not permitted to be used to supply water for domestic purposes, so manufacturers do not necessarily need to be quite so fussy about the materials that they are made from.

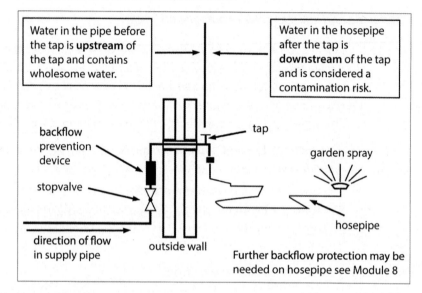

However, if any of these pipes and components are to be connected to a pipe that also supplies drinking water, the drinking water supply will need to be protected from any possible contamination that these materials might cause. This of course, is done using backflow prevention devices. (Backflow is dealt with in more detail in Module 8.)

Paragraph 2 of Schedule 2 is about materials and substances in contact with water.

So! What is important about the materials we use, and what should you know about them?

Any water pipe or fitting that carries water for domestic use or for food preparation, whether it is a supply pipe or a distributing pipe, must be manufactured from a material that is not likely to contaminate the water.

There are a variety of regulations and standards governing the use of materials in the manufacture of water fittings. For example, the Guidance document suggests that non-metallic materials will be acceptable if they comply with BS 6920-1:2014.

Manufacturers will need to know this! But, what do you, the installer need to know?

How can I be sure that the materials I am purchasing are approved for use and will not cause any contamination?

Firstly, you could look to see if the material is covered by a British or European Standard Specification.

Where a product is certificated to a BS or EN Standard the manufacturer will have taken his product through a series of tests which are regularly checked under BSI or EC Quality Control schemes.

The manufacturer will want you to know that his product is to a recognised quality standard, and either the product or its packaging will be clearly marked.

It should be remembered however that British Standards Specifications are guidance documents and have no force of law **unless referred to specifically by the Regulations**.

Secondly, you should always be prepared to **take advice**, and one organisation to get advice **from**, is of course, **your local water undertaker**. Another important organisation is the **Water Regulations Advisory Scheme (WRAS)**.

Formerly known as the Water Byelaws Scheme, the **WRAS** has been carrying out fittings testing for many years and will continue to advise on Water Regulations in the future. As part of their work they produce a **Fittings and Materials Directory** which lists all approved fittings. This Directory is an important guide to all who aim to comply with or enforce water regulations.

When purchasing materials, look for these signs on the product or the packaging.

The **BSI Kitemark** along with the BS number shows that the product has been fully tested under BSI's quality testing scheme.	The WRAS approved product symbol shows that the product has been tested for approval and is listed in the Fittings and Materials Directory.	The **WBS teardrop symbol**, and still shown on some products tested before July 1999.	The **CE mark** indicates that the product has been tested to EN Standards and may legally be placed on the market.

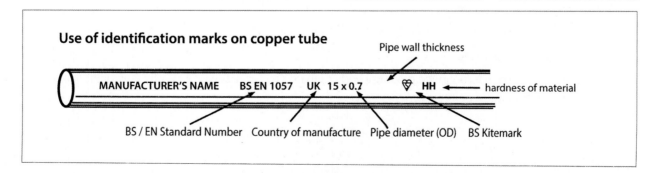

Use of identification marks on copper tube

What else should I know about when selecting materials?

We have already mentioned two materials referred to in the Guidance document as health hazards and these are specifically banned from use in potable water systems.

The two banned materials are:

1. Lead and any material or substance containing lead.

2. Any bituminous coatings produced from coal tar.

There are many existing installations containing lead pipes, and lead could be a hazardous material. So! What should you do if you come up against a lead pipe and have to make a connection to it?

Firstly, customers should be advised of the need to remove as much of the lead pipe as possible and preferably to remove all of it!

Secondly, where a connection must be made to lead pipe, it should be done in a way that will not lead to any further contamination.

Use only fittings that are manufactured for the purpose and install in a way that galvanic (electrolytic) corrosion cannot take place. Do not solder copper tube directly into the lead pipe, and never install any copper tube so that the water passes through the copper tube before passing on through the lead pipe. Why? Because a direct connection between copper and lead will most likely result in electrolytic action, causing the lead to be dissolved and taken into solution to contaminate the water.

Water passing through copper will dissolve some of the copper from the pipe which will encourage corrosion in any lead pipe that it passes through, which again may be taken into solution as the water passes through.

Materials should be chosen with care and consideration given to the purpose for which they are to be used.

Factors that may affect the selection of materials are listed in the Guidance document.

These include:

(a) The effects of temperature, internally and externally on the material or fitting.

   e.g. Will the material soften with heat or become brittle when cold?

(b) The effects of corrosion and the compatibility of metals,

   e.g. Is the material likely to become corroded by the water or if underground, by the soil conditions?

   Are different materials likely to react with one another to cause corrosion or deterioration in material quality?

(c) Aging, fatigue, and durability

   e.g. How long are the materials expected to last in good condition?

(d) The permeability of materials.

   e.g. Will the gases pass through the pipe wall and are there any other services nearby that might affect the water supply through permeation?

Jointing materials

Some jointing materials have been given individual consideration and are mentioned specifically in the Guidance document. These are as follows:

Soft solders used **for jointing copper or copper alloy tubes** should comply with BS EN 29453 now superseded by BS EN ISO 9453:2006 and may be of two types:

- tin/copper alloy Type No. 23 or 24, or
- tin/silver alloy Type No. 28 or 29.

Lead solders are a particular problem. The Regulations follow previous byelaws which prohibited the use of lead solders on pipework used for domestic purposes (e.g. hot and cold water supplies) although their use was still permitted on installations such as central heating systems where the water is not drawn off.

However, despite the ban, there are still many recently reported cases of water contamination through the use of lead solders on copper tubes. In many cases the operator carries two types of solder for the two separate installations, but it is easy to use the wrong one!

The answer is to use lead free solders for all installations.

Silver solder or silver brazing metal used for the jointing of copper and copper alloy tubes should comply with: BS 1845 now superseded by BS EN 1044:1999: Table 2 types AG14 or AG20. Copper phosphorus brazing filler metals should comply with BS 1845 Table 3 and be of types CP1 to CP6. For safety reasons all brazing alloys should be cadmium free.

Jointing compounds used for sealing screw threaded joints should comply with BS 6956: Part 5.

PTFE tape (unsintered polytetrafluoroethylene) should comply with:
- BS 6974 now superseded by BS EN 751-3:1997 for thread sealing applications, and
- the material with BS 6920: Part 1 now superseded by BS 6920-1:2014.

Remember!

It is important that **any material you use**:

i) **complies with a relevant British or European Standard Specification**, particularly those that are referred to in the Regulations or

ii) **is listed in the Water Fittings and Materials Directory** produced by the Water Regulations Advisory Scheme, or

iii) both. Many materials will be covered in both ways

Note: Many European Standards will have an identical British Standard number.

e.g. *BS EN 1057 'Copper and copper alloys – seamless round copper tubes for water and gas in sanitary and heating applications'.*

Self-assessment questions

1. Compete the following statement.

   a) *The purpose of Paragraph 2 of Schedule 2 is to ensure that materials we use will not cause any* _____ *that will affect the drinking qualities of the water.*

2. Paragraph 2 is concerned with fittings used for the supply of water for domestic and food production purposes.

   Explain briefly what is meant by 'Water for Domestic Purposes'.

   _____

   _____

   _____

3. State the TWO ways in which we can identify fittings that are approved for use under sub-paragraph 2.(1) of Schedule 2.

   1. _____

   2. _____

4. Water fittings such as hosepipes, flushing cisterns, feed and expansion cisterns and closed circuits *'downstream of a terminal fitting supplying wholesome water'* are exempted from the requirement not to contaminate water. But there are TWO conditions.

   What are the TWO conditions under Paragraph 2(2) that are applied to these fittings?

   1. _____

   2. _____

5. British Standard Specifications are 'advisory' documents. State the conditions that may cause a BS to become a requirement under Water Regulations.

   _____

   _____

   _____

6. Give the name of the Organisation that gives guidance on Water Regulations and is responsible for maintaining a 'list of approved fittings'.

   _____

   _____

7. Identify each of the symbols shown below

1. _____

_____

2. _____

_____

3. _____

_____

4. _____

_____

8. State TWO materials that are specifically banned under Schedule 2

1. _____

2. _____

Check your answers on pages 11 and 12.

Summary of main points:

- **The purpose of Paragraph 2 of Schedule 2 is to prevent contamination of water by contact with materials.**

- Materials used should not cause the water to become toxic, or biologically unhealthy, or to affect its colour, taste, or odour, or alter the composition of the water supplied by the water undertaker.

- **Paragraph 2 is concerned with fittings used for the supply of water for 'domestic and food production purposes'.**

  *Water for domestic purposes means '**wholesome water supplied by a water undertaker for general use including drinking'.***

- Paragraph 2 does not cover water for non-domestic use i.e. hosepipes, flushing cisterns, feed and expansion cisterns, closed circuits and warning pipes, provided the water supply is fully protected from risk of backflow.

- Materials covered by a British Standard Specification give a good indication of their suitability, **but British Standard Specifications are advisory documents and have no force of law unless referred to specifically by the Regulations.**

- There may be a requirement to comply with certain EN Standards through European law.

- **Two materials are banned under Schedule 2 are:**

  1. Lead and any material or substance containing lead.
  2. Any bituminous coatings produced from coal tar.

- **Use only materials that are manufactured for the purpose, and install in a way that electrolytic corrosion cannot take place.**

- **Advice on water fittings and materials can be obtained from the Water Regulations Advisory Scheme (WRAS).**

- The Water Regulations Advisory Scheme produces a **Fittings and Materials Directory** which **lists approved fittings.**

Note: The Water Regulations Advisory Scheme also produce a Water Regulations **Guide to the Water Regulations** which is essentially an update of the well known and established 'Water Supply Byelaws Guide'.

Answers to self-assessment questions

1. Complete the following statement.

   a) *The purpose of Paragraph 2 of Schedule 2 is to ensure that materials we use will not cause any* **contamination** *that will affect the drinking qualities of the water.*

2. Paragraph 2 is concerned with fittings used for the supply of water for domestic and food production purposes.

   Explain briefly what is meant by 'Water for Domestic Purposes'.

   Water for domestic purposes means **'water of drinking quality supplied by a undertaker for general use including drinking'.**

3. State the TWO ways in which we can identify fittings and components that are approved for use under sub-paragraph 2.(1) of Schedule 2.

   1. **We can check if they are certificated by a BS or EN Standard.**

   2. **We can check to see if they are listed in the Materials and Fittings Directory.**

4. Water fittings such as hosepipes, flushing cisterns, feed and expansion cisterns and closed circuits *'downstream of a terminal fitting supplying wholesome water'* are exempted from the requirement not to contaminate water. But there are TWO conditions.

   What are the TWO conditions under Paragraph 2(2) that are applied to these fittings?

   1. **the use to which the water down stream is put does not require wholesome water**

   2. **a suitable arrangement to prevent backflow is installed**

5. British Standard Specifications are 'advisory' documents. State the conditions that may cause a BS to become a requirement under the Water Regulations.

   A British Standard Specification becomes law only if it is referred to in the Regulations.

6. Give the name of the Organisation that gives guidance on Water Regulations and is responsible for maintaining a 'list of approved fittings'.

   The Water Regulations Advisory Scheme *is approved to keep the list of fittings*

7. Identify each of the symbols shown below

1. *The British Standards Kitemark*

2. *The WRAS approved fittings symbol*

3. *WBAS 'teardrop'(approved fittings) symbol*

4. *The CE (European Product Standard) symbol*

8. State TWO materials that are specifically banned under Schedule 2.

1. *Lead and any material containing lead*

2. *Any bituminous coatings produced from coal tar*

What to do next

Well done!

I hope that you have found Module 3 interesting.

I suggest you now go on to Requirements for water fittings **Module 4**

Water Industry Act 1991:

Water Supply (Water Fittings) Regulations 1999

An Open Learning Course

Module 4

Requirements for water fittings

Introduction

This Module looks at **Paragraphs 3 to 7** of Schedule 2 which are **concerned with how water fittings are used**.

We looked at water fittings in Module 3 in respect of the materials and substances that water fittings were made of. In Module 4 we take this a stage further by studying the way we use water fittings **so that contamination or waste of water will not occur** as a result of the water fittings we install.

In this module we enlarge a little further on your responsibility (outlined in Module 3) for the materials you use in your installation work.

There are all sorts of reasons why water might become contaminated or wasted as a result of the type of fitting we use, or by the way we use them. You will see, as we go through, that Paragraph 3 sets the scene and prescribes that water fittings should not be affected by galvanic action, or damage from other causes such as external load, stress or settlement, pressure surges, or temperature fluctuations.

Paragraph 4 goes a little further and prescribes that water fittings should be watertight, and not permit ingress by contaminants, not be damaged by frost, or substances that might cause the material to deteriorate. It also looks at support for water fittings.

Paragraph 5 is concerned with the ability of water fittings to resist internal pressures whilst Paragraph 6 says that fittings installed should not affect the quality of the water or the pressure in the main.

Finally Paragraph 7 looks at 'concealed water fittings' and sets out a number of additional precautions to further protect those fittings that cannot be seen. It also gives a definition so we are in no doubt as to what is meant by 'concealed water fittings'.

What is the requirement?

Schedule 2: Paragraphs 3 to 7: requirements for water fittings

3. Every water fitting shall:

    (a) be immune to or protected from corrosion by galvanic action or by any other process which is likely to result in contamination or waste of water; and

    (b) be constructed of materials of such strength and thickness as to resist damage from any external load, vibration, stress or settlement, pressure surges, or temperature fluctuation to which it is likely to be subjected.

4. Every water fitting shall:

    (a) be watertight

    (b) be so constructed as to:

        (i) prevent ingress by contaminants, and

        (ii) inhibit damage by freezing or any other cause;

    (c) be so installed as to minimise risk of permeation by, or deterioration from contact with, any substance which may cause contamination; and

    (d) be adequately supported

5. Every water fitting shall be capable of withstanding an internal water pressure not less than 1½ times the maximum pressure to which that fitting is designed to be subjected in operation.

6. No water fitting shall be installed, connected or used which is likely to have a detrimental effect on the quality or pressure of water in a water main or other pipe of a water undertaker.

7. **(1)** No water fitting shall be embedded in any wall or solid floor.

    (2) No fitting which is designed to be operated or maintained, whether modulely or electronically, or which consists of a joint, shall be a concealed water fitting.

    (3) Any concealed water fitting or mechanical backflow device, not being a terminal fitting shall be made of gunmetal, or another material resistant to dezincification.

    (4) Any water fitting laid below ground level shall have a depth of cover sufficient to prevent water freezing in the fitting.

    (5) In this paragraph 'concealed water fitting' means a water fitting which:

        (a) is installed below ground;

        (b) passes through or under any wall, footing or foundation;

        (c) is enclosed in any chase or duct; or

        (d) is in any other position which is inaccessible or renders access difficult

---

Firstly let's look again at water fittings. The whole of this section, Paragraphs 3 to 7 is about water fittings.

What are water fittings?

Within the Regulations, 'water fittings' is a general term given to a variety of separate fittings and includes pipes, taps, valves, meters, cisterns and cylinders, baths, water closets and other sanitary appliances, boilers and hot water storage vessels, washing machines, etc. In fact, if any item within a premises contains or uses water supplied by the water undertaker, it is a water fitting.

What does Paragraph 3 say about corrosion and waste of water in respect of water fittings?

Sub-paragraph 3(a) requires **water fittings to be immune to or protected from galvanic action**. Galvanic action is a term associated with the **electrolytic corrosion of dissimilar metals in a damp or wet environment**. The rate at which electrolytic corrosion will occur depends on its position in the 'electromotive series'. The further apart the metal is on the chart the more likely corrosion will occur and when two metals are placed in contact with each other the metal at the base end of the scale will be the one to corrode.

A common example is seen in the corrosion of galvanised steel cisterns when connected to copper pipework systems. Look at the electro-chemical series below and you can see that copper and zinc are some distance apart and zinc being the base metal, will corrode.

Metal	Chemical symbol	Electrod potential (volts)	
Silver	Ag	+0.80	Cathodic protected end
Copper	Cu	+0.35	(noble end')
Hydrogen	H	0.00	
Lead	Pb	- 0.12	
Tin	Sn	- 0.14	
Nickel	Ni	- 0.23	
Iron	Fe	- 0.44	
Chromium	Cr	- 0.56	
Zinc	Zn	- 0.76	
Aluminium	Al	- 1.00	
Magnesium	Mg	- 2.00	
Sodium	Na	- 2.71	Anodic, corroded end (base end)

The electro-chemical series

The further apart the materials are on the scale, the more the electrical potential, and the greater the risk of corrosion. It is therefore important that materials be chosen which great care to avoid this corrosion risk.

Another example! Remember what we said in Module 3 about not connecting copper pipe directly into lead pipe? Two things are likely to happen when the lead corrodes:

1. lead will be taken into solution to contaminate the water, and

2. the lead will be weakened by the corrosion which may result in leakage.

In sub-paragraph 3(b) **water fittings** are required to **resist the effects of external load, vibration, stress or settlement, pressure surges, or temperature fluctuations** to which they are likely to be subjected.

Most fittings and appliances are subjected to stresses. For instance, hot water components are continually heated and cooled, and pipes run through buildings and under roads and driveways that expand and contract and settle or vibrate. Sometimes they are subjected to cold and frost. All of these things put stresses on the materials which water fittings in service must be able to resist, and remain watertight [Para. 4(a)]. However! It is not enough to rely on the fittings themselves to be watertight, we also have to **select and install our water fittings so that any stress is kept to a minimum.** For example, pipes passing through walls should be sleeved or ducted to reduce stress caused by movement of the wall or expansion of the pipe. Paragraphs 4 to 7 takes this theme a step further.

So! How do we know if the water fittings we select comply with the requirements of Paragraph 3(b)? Again think back to what was discussed in Module 2. **Water fittings must be adequate for their purpose, and comply with an appropriate standard**, for example a BS or EN Specification.

Now to Paragraph 4

Sub-paragraph 4(a) requires water fittings to be watertight. Quite a simple statement really, but as we all know this is not as easy as it seems. To be sure they are watertight means using proper jointing techniques that will **remain watertight during service**, making all the proper commissioning checks to ensure all fittings are sound, and it means testing to the required standards (see Module 6).

This requirement is supported by **Paragraph 5** in which water fittings are required to be **capable of withstanding 1½ times the expected maximum internal pressure** that they are designed to be subjected in operation. Much of this requirement is down to the manufacturer, but again you have a responsibility to see that fittings are used correctly and within the limits laid down in manufacturers' literature.

A good example of this is in the use of copper cylinders that are graded to suit various pressure conditions. A grade 4 copper cylinder to BS 1566-1:2002 +A1:2011 is suitable for pressures of up to 6m head. This means that it must be able to withstand a test pressure of 9m head and the manufacturer will test at even higher pressures. As far as installation is concerned it is no good fitting a grade 4 copper cylinder where it could be subjected to more than 6m head because it will not be adequate for its purpose and as a result you will have no valid claim under the manufacturer's guarantee.

Typical cylinder maximum pressures								
Type and BS number	grade	max. head	grade	max. head	grade	max. head	grade	max. head
Galvanised steel cylinders								
– cylinder – direct to BS 417: Part 2: 1987	A	30m	B	18m	C	9m	n/a	n/a
and indirect to BS 417: Part 2: 1987	1	25m	2	15m	3	10m	n/a	n/a
Copper cylinders								
– direct to BS 1566-1:2002 + A1:2011	1	25m	2	15m	3	10m	4	6m
– double feed indirect to BS 1566: Part 1: 2002+A1:2002	1	25m	2	15m	3	10m	4	6m
– single feed indirect to BS 1566: Part 2: 1984	n/a	n/a	2	15m	3	10m	4	6m

Paragraph 4(b) covers much more ground, looking to prevent ingress of contamination in fittings and the prevention of damage by freezing or other causes. We will take these three items separately.

What is meant by ingress of contaminants and how can I deal with it?

Sub-paragraph 4(b)(i) says *'Every water fitting shall be so constructed and installed as to prevent ingress by contaminants'* which means that **water fittings** are required **not to be placed in such a position that impurities can get into water fittings to contaminate the water** inside. So! Let's have a look at some examples.

Water storage cisterns are required **to have rigid, close fitting, and securely fixed cover and be fitted with screened overflows and vents** to prevent insects and dust from entering the cistern. The so called Byelaw 30 cistern (now Schedule 2, Paragraph 16) is still as important now as it was under the Byelaws that led to its introduction. *(More on cisterns in Module 9)*

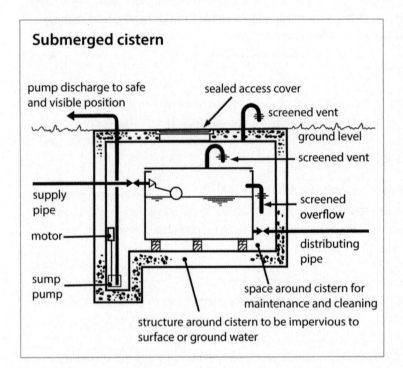

Submerged cistern

pump discharge to safe and visible position

sealed access cover

screened vent

ground level

screened vent

supply pipe

motor

screened overflow

distributing pipe

sump pump

space around cistern for maintenance and cleaning

structure around cistern to be impervious to surface or ground water

Water storage cisterns should not be placed in positions whereby they could become flooded by rainwater or ground water or by the cistern overflowing.

Where it is necessary for a cistern to be in such a position, it should be **installed in a watertight enclosure** which could be a concrete chamber or basement that is watertight.

In either case the chamber or basement should be **fitted with an electrical sump pump** to remove any water that might collect in the base of the chamber. **Audible or visual devices** should be fitted to **show if the cistern reaches overflowing level**, and to show if any water is building up in the base of the chamber.

Cisterns, cylinders and other components should be arranged so that they cannot become contaminated by the build-up of sediment. If necessary filters should be fitted to prevent sediment from the insides of pipes passing along and into the system. Regular maintenance and cleaning is essential to prevent undue build-up of sediment which will provide an ideal environment for bacterial growth particularly where water is likely to be warmed.

Draining valves should be positioned where they cannot become submerged so draincocks in sump holes below floors are not permitted nor are draincocks below ground. In the latter case they should be fitted above ground but as low as is reasonably practical. (More on draining valves in Paragraph 11, see Module 5)

Draining valve on drops below floor

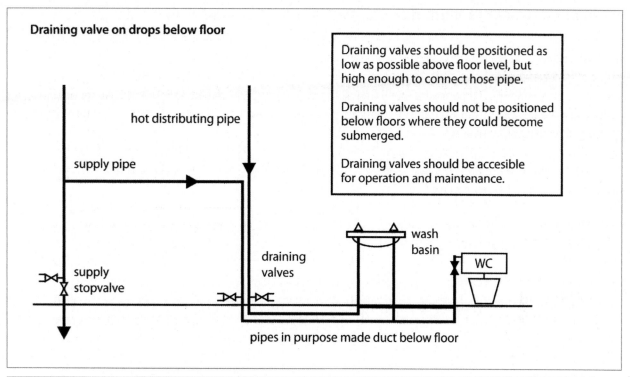

Draining valves should be positioned as low as possible above floor level, but high enough to connect hose pipe.

Draining valves should not be positioned below floors where they could become submerged.

Draining valves should be accesible for operation and maintenance.

Draining valve on garden stand/pipe

Draining valves should be positioned as low as possible above ground level, but high enough to drain stand-pipe without draincock becoming submerged.

Draining valves NOT to be positioned below ground level.

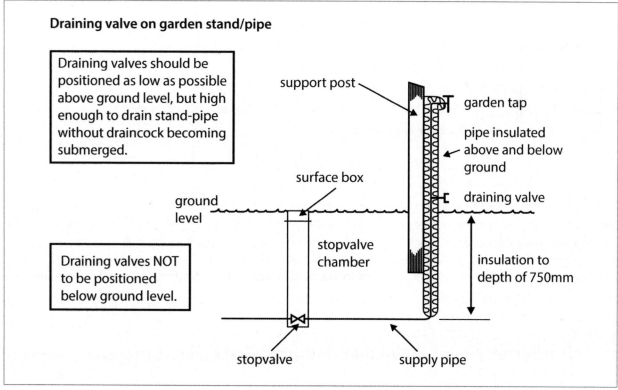

What about frost precautions?

Frost precautions are the subject of sub-paragraph 4(b)(ii) which says *'Every water fitting shall be so constructed and installed as to inhibit damage by freezing or any other cause'*.

This is a simple statement that covers pretty well all the frost precautions we need to apply to a hot or cold water installation. It also covers, as you can see, damage by causes other than freezing. We'll come to that later but now a look at frost precautions.

The best source of information on frost precautions is probably BS 8558/BS EN 806-4 which some time ago absorbed the contents of the earlier BS 6700 'Frost precautions'.

We are all aware of the problems associated with frost damage so we will simply concentrate on what is needed to comply with sub-paragraph 4(b)(ii).

The first thought for most people when considering frost is insulation, but, as important as insulation is, it is not the entire answer to the frost problem. No matter how much insulation we use, it will only delay the onset of frost. In most cases the delay will prevent loss of water or damage. but in severe weather any insulation barrier will eventually be breached unless other measures are also taken.

So! What should we do?

Firstly we can **avoid** installing pipes and fittings in **areas** that are known to be difficult to keep warm, **such as**:

- **unheated parts of the building,** e.g. roof spaces, cellars, garages and outhouses;
- **known draughty areas** within buildings e.g. near doors, windows or ventilation openings, in 'cold' roofs, spaces and beneath suspended ventilated floors;
- **cold surfaces** e.g. direct contact with outside walls or ducts or chases within outside walls;
- **any position outside buildings and above ground**.

So much for what we **should not do**, now on to what we **should do**!

Underground pipes and fittings are the subject of sub-paragraph 7(4) where it states that *'any water fitting laid below ground level shall have a depth of cover sufficient to prevent water freezing in the fitting'*. Nothing new here! The **required depth of cover is**, as it has been for a long long time, **not less than 750mm** for frost protection. Not many frosts in this country get down to a depth of 750mm.

Incidentally, this depth will also provide protection from 'other causes' such as garden digging, ground movement, vibration due to traffic and so on.

Additionally **fittings below ground should not be more than 1350mm deep**, otherwise they become inaccessible for maintenance and repair.

When laying pipelines in uneven ground, care should be taken to ensure that the minimum depth of cover is maintained over the whole length of the pipeline.

This is especially important on new site installations where the ground is often levelled after the pipeline has been laid.

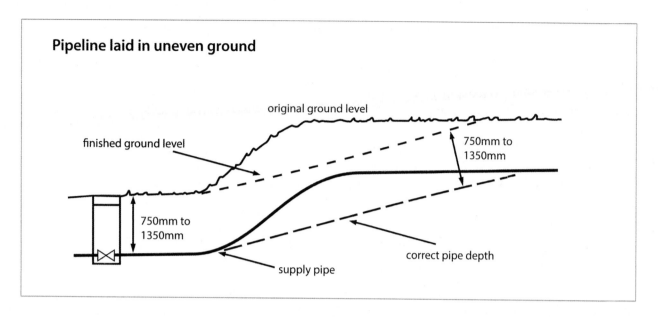

Pipeline laid in uneven ground

original ground level

finished ground level

750mm to 1350mm

750mm to 1350mm

correct pipe depth

supply pipe

There will be occasions **when it is difficult or impossible to achieve the correct depths of between 750mm and 1350mm**. In these cases, remember, you are required under regulation 5 **to notify the water undertaker of the work**.

The following two diagrams illustrate ways to overcome obstructions to the pipeline that might cause the pipeline to deviate from the prescribed depth.

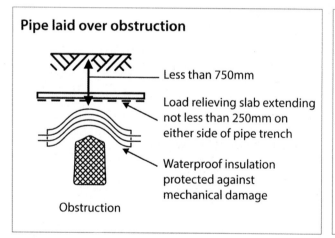

Pipe laid over obstruction

Less than 750mm

Load relieving slab extending not less than 250mm on either side of pipe trench

Waterproof insulation protected against mechanical damage

Obstruction

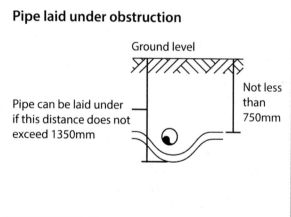

Pipe laid under obstruction

Ground level

Pipe can be laid under if this distance does not exceed 1350mm

Not less than 750mm

What do I do with pipes that are rising up from below ground to serve a fitting outside?

Pipes rising from below ground to fittings above ground level **should be fitted with adequate water proofed insulation that extends to a depth of 750mm**. down into the ground **and which is protected from 'mechanical or other damage' above ground**. The diagram below shows a cattle trough.

The situation is similar for outside taps that are fed from within a building.

Pipe rising from below ground to cattle trough

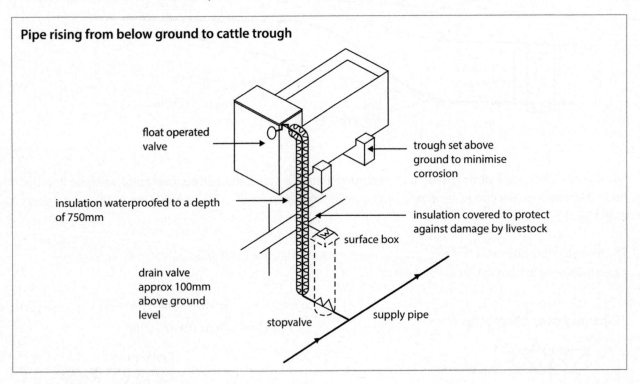

Insulation of outside taps

a) tap fed from within building   **b) tap fed from below ground**

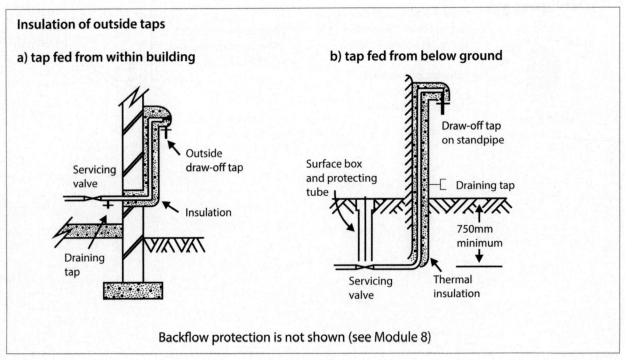

Backflow protection is not shown (see Module 8)

Is there a problem with supply pipes entering buildings?

Generally, **pipes entering buildings should**:

– **be laid in a duct** that will allow access for renewal and repair and also give some degree of insulation

– **enter through or under the foundations** at a minimum **depth of 750mm**

– **rise up** in the building **at least 750mm from the outside wall** (if less than 750mm the pipe should be insulated); and

– **where the pipe rises up through a suspended** (ventilated) **floor,** the pipe should **be insulated** from floor level **to a depth of 750mm** below ground level.

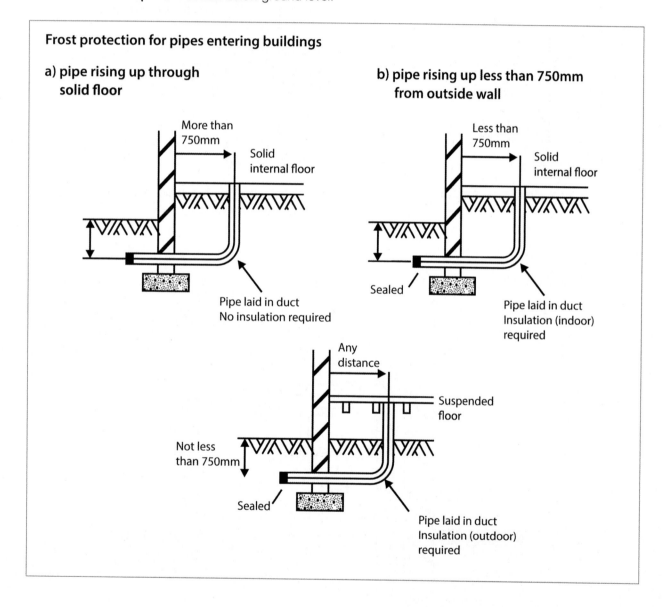

Frost protection for pipes entering buildings

a) pipe rising up through solid floor

b) pipe rising up less than 750mm from outside wall

How should I deal with pipes and fittings near cold surfaces or in unheated positions?

First let us look at pipes and fittings near to or against outside walls. **Pipes and fittings should NOT be chased or ducted into outside walls. They should be spaced away from the wall** using spacer clips or if saddle clips are used, to a base board that will provide insulation against cold from the wall.

Clipping to pipes on outside walls

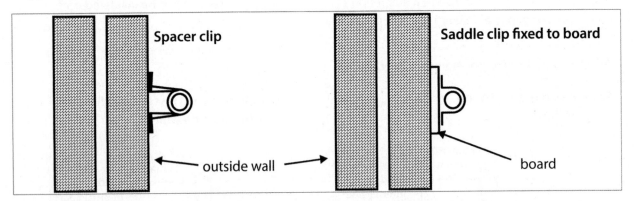

All water fittings within unheated places such as garages and outbuildings **should be insulated** and if the pipe rises from below the floor, the insulation should be taken down to a depth of 750mm.

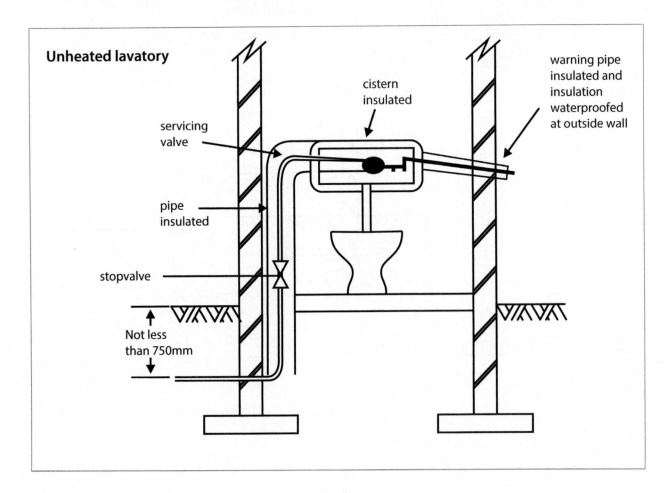

What should I know about frost precautions for water fittings in roof spaces?

All water fittings in roof spaces, including cisterns, supply pipes, distributing pipes, and vent pipes are required to be insulated.

Cold roofs. Where the roof space is fitted with **loft insulation at ceiling level, (cold roof) the ceiling insulation should be** omitted beneath cisterns so as to allow warmth from the room below to reach the cistern, and **where possible pipes** should be **kept beneath the ceiling insulation.**

Warm roofs. (Those having insulation near the surface of the roof thus keeping the roof space warm). Whilst there is no need to insulate against frost in a 'warm' roof, insulation should still be provided to all cisterns pipes and fittings to prevent the water from being unduly warmed.

It is advisable to keep all water fittings, pipes and cisterns as far from the roof surface as possible.

Cistern in roof space

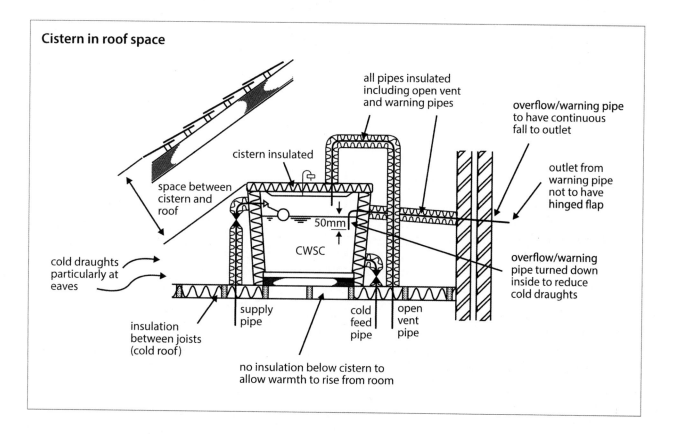

Are there any problems I should be aware of for water fittings above ground outside of buildings?

Yes there are! It should be born in mind that pipes and fittings outside of buildings are extremely vulnerable to frost and to the effects of wind and draughts around trees, bushes and buildings. **Do NOT position fittings outside buildings if it can be avoided.**

Where pipes and fittings are positioned outside buildings insulation is essential and the insulation **must be waterproofed. Wet insulation has virtually no insulation value.** Pipes and fittings must also be protected from mechanical and other damage.

The following two examples illustrate outside pipe installations.

Pipe passing between buildings

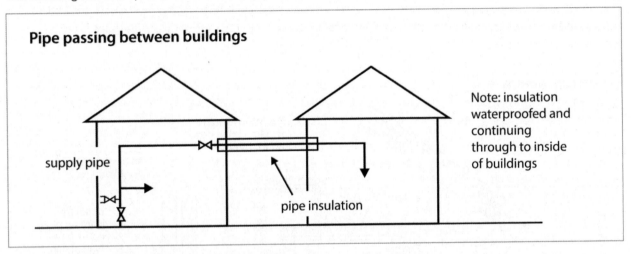

supply pipe

pipe insulation

Note: insulation waterproofed and continuing through to inside of buildings

Pipe rising to caravan

Note: Increase in wind speed around and under caravan increases the chill factor and makes pipe extra vulnerable to frost damage

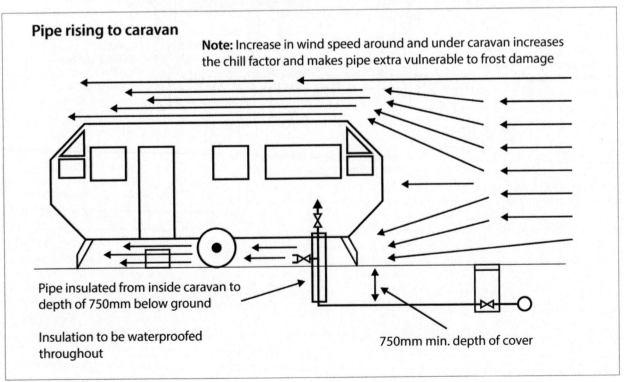

Pipe insulated from inside caravan to depth of 750mm below ground

Insulation to be waterproofed throughout

750mm min. depth of cover

What insulation should I provide?

Insulation is quite a complicated subject. BS 5422-2009 provides a number of tables of calculated insulation thicknesses for pipework in a variety of situations, all of which should be read with care because the figures given are in some cases not economically viable. The Guidance to Schedule 2, G4 23, discusses the background to the criteria used in BS 5422-2009 for calculating insulation, and considers two conditions, normal and extreme, where differing thicknesses of insulation might apply.

Normal conditions would include water fittings installed within buildings in areas that are not subjected to draughts from outside the building. Examples given are cloakrooms, store rooms, utility rooms, roof spaces where pipes are below the ceiling insulation, and unheated parts of otherwise heated commercial buildings.

Extreme conditions include water fittings installed inside unheated or marginally heated buildings, or below suspended floors, cold roof spaces or other areas where draughty conditions are likely. Exposed conditions will obviously include water fittings installed outside of buildings.

Unfortunately Table 4.2 from the Guidance Document (repeated below) only gives figures for normal exposures in commercial buildings and the guidance suggests that an increased insulation value should be applied to extreme conditions, and that the advice of insulation manufacturers should be sought.

Table 4.2 from the Guidance notes to Schedule 2.

Table 4.2: Recommended minimum commercial thicknesses of thermal insulation for copper water pipes of minimum wall thickness complying with BS EN 1057 in normal conditions of exposure.										
External diameter of pipe	Thermal conductivity of insulation material at 0°C in W/m.K)									
	0.02		0.025		0.03		0.035		0.04	
mm	mm		mm		mm		mm		mm	
15	20	(20)	30	(30)	25*	(45)	25*	(70)	32*	(91)
22	15	(9)	15	(12)	19	(15)	19	(19)	25	(24)
28	15	(6)	15	(8)	13	(10)	19	(12)	33	(14)
35	15	(4)	15	(6)	9	(7)	9	(8)	13	(10)
42 and over	15	(3)	15	(5)	9	(5)	9	(5)	9	(8)

The Table refers to thermal conductivity which is the rate at which heat will pass through a material. There is a variety of insulation materials available, each of which has a different resistance to the passage of heat and therefore a different thermal conductivity value. This is reflected in Table 4.2. For further information readers might also refer to the guidance document and notes that go with Table 4.2.

Are draining valves considered to be a frost protection measure?

Yes they are, although the Regulations tend to look at them more as a means of draining down for repair and this is looked at in Module 5.

Draincocks are perhaps most useful when systems are out of use for a while and the building is unheated, particularly in the winter. A typical example is when holidays are taken during the winter and the house is empty. In these cases pipes and fittings should be drained down for the period that the premises are not in use. Empty pipes cannot suffer from frost damage.

Draincocks to BS 2879 should be positioned on all low points in the system so that the entire system can be emptied. The only exception to this is in positions mentioned earlier on page 6 where a draincock on the lowest point in the system might be a contamination risk.

Finally! Let's go back to where we began with insulation on page 8.

No amount of insulation can be effective in protecting against frost over long periods of extreme weather conditions. **The best protection is to maintain temperatures of rooms containing water fittings to a level that frost cannot occur.** This can be achieved by space heating of the building or localised heating in the immediate area that the water fittings are located.

One such localised method that is acceptable is the use of self regulating trace heating to BS EN 60079-30-2:2007. However, a nominal thickness of insulation should still be provided to guard against failure of the heating used.

What is meant by permeation or deterioration?

Sub-paragraph 4(c) **risk of permeation or deterioration by contact with substances that may cause contamination should be minimised**.

Some plastics materials may be damaged by contact with certain other fluids and gases. Plastics pipes and fittings when exposed to petrol and oil for instance may penetrate, soften, and weaken the pipe material, which may become contaminated as a result. Gases and fumes have been known to permeate (pass through) the walls of plastic pipes without any apparent damage to the pipe but introducing a smell or taste into the water.

Where it is known that contamination from oil, petrol (hydrocarbons) or other substances could possibly occur, other materials should be used that are resistant to permeation.

An example is seen in the supply of water to outlets at petrol filling stations, where copper tube would be much preferred to plastics.

Now what about adequate support?

Every water fitting shall be adequately supported. [Sub-paragraph 4(d)]

Here we should remember that 'water fittings' means pipes, joints, valves, cisterns, cylinders, and any other component that carries water from the water undertaker's main.

Pipes should be securely fixed using appropriate clips or brackets **and spaced in accordance with the recommendations of BS 8558/BS EN 806-4. Allowance should be made** within the clips for thermal movement to take place. Inadequate fixings may lead to damage to pipes particularly those of plastics materials that may sag between clips that are spaced too far apart. Inadequate fixings may encourage noise and vibration in pipework systems that is both a nuisance and risk of damage to the pipes. Restriction of pipes within clips could lead to stress damage through thermal expansion.

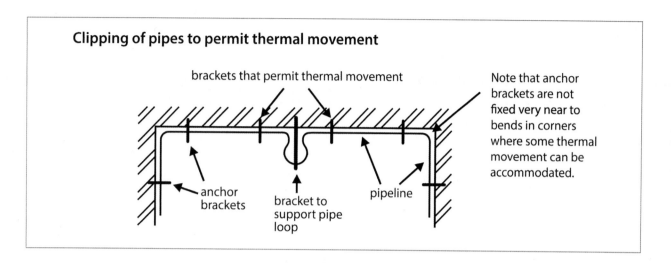

Clipping of pipes to permit thermal movement

brackets that permit thermal movement

Note that anchor brackets are not fixed very near to bends in corners where some thermal movement can be accommodated.

anchor brackets

bracket to support pipe loop

pipeline

Pipes in ducts and roof spaces are often missed and should receive special attention. Pipes connected to cisterns and cylinders need support, otherwise their weight is likely to cause stress on joints. This is particularly important where connections are made to plastic cisterns. And don't forget overflow pipes and warning pipes. These are rarely properly clipped.

Cold water storage cisterns need proper support. They carry a great deal of weight which should be spread across as many ceiling rafters as possible. **Flexible cisterns should be supported on a continuous platform that supports the whole of the base of the cistern.**

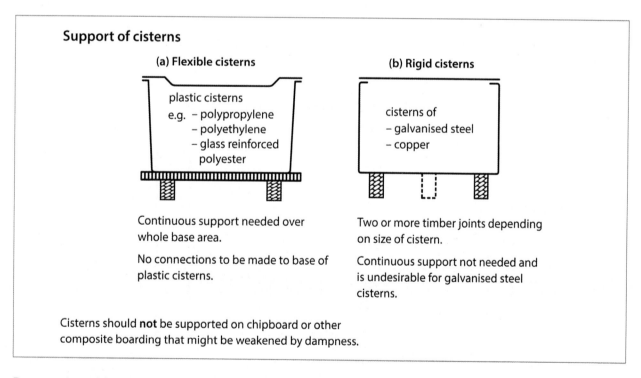

Support of cisterns

(a) Flexible cisterns

plastic cisterns
e.g. – polypropylene
– polyethylene
– glass reinforced polyester

Continuous support needed over whole base area.

No connections to be made to base of plastic cisterns.

(b) Rigid cisterns

cisterns of
– galvanised steel
– copper

Two or more timber joints depending on size of cistern.

Continuous support not needed and is undesirable for galvanised steel cisterns.

Cisterns should **not** be supported on chipboard or other composite boarding that might be weakened by dampness.

Paragraph 6 of Schedule 2 says that water fittings are required not to have a detrimental effect on the pressure or quality of water in a main or other pipe of the water undertaker.

So! To quality of water first!

A large percentage of the Regulations deal with contamination and preservation of water quality in one form or another. So there is no need to repeat it all here.

Let's just simply say at this stage that any water in a pipe that has been supplied by a water undertaker is required to be wholesome. Your duty as an installer is to make sure that nothing you do will allow any water that has been used, or in any way contaminated, can return to the main.

The other modules and in particular, Modules 7 and 8 will take this further.

Now to pressure in the main.

How can installations affect water pressure?

The most obvious installation that might affect water pressure is a booster pump. Two points here!

1. **It is usual for water undertakers to insist on a break cistern to be fitted to separate the booster pump from the incoming supply pipe.** This will prevent large quantities of water from being drawn off that might reduce the pressure in the main and cause loss of pressure and reduce the amount of water available to other customers.

2. Remember what was said in Module 2. Regulation 5 says we must **notify the water undertaker of any pump or booster installation that draws more than 12 litres per minute from the supply pipe.** By notifying the water undertaker of the installation he will then have the opportunity to make sure that the installation will not have an adverse effect on the water pressure in the main.

The diagram below shows the installation of a break cistern for a boosted water supply.

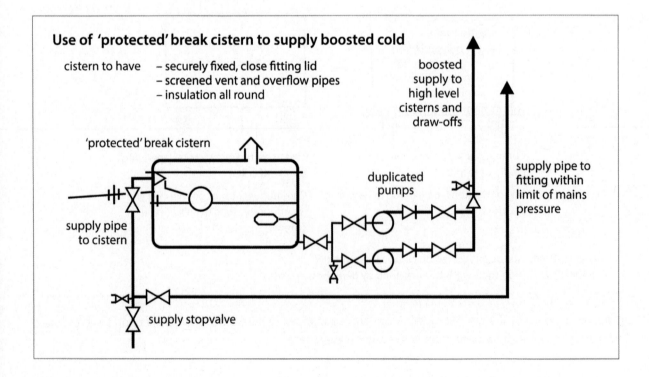

Use of 'protected' break cistern to supply boosted cold

cistern to have — securely fixed, close fitting lid
— screened vent and overflow pipes
— insulation all round

boosted supply to high level cisterns and draw-offs

'protected' break cistern

supply pipe to cistern

duplicated pumps

supply pipe to fitting within limit of mains pressure

supply stopvalve

How should I deal with concealed water fittings?

This is the subject of Paragraph 7 of Schedule 2.

The rules for concealed water fittings are relatively simple, and are given in sub-paragraphs 1 to 4.

1. **No water fitting shall be embedded in any wall or solid floor.** This means that chasing pipes into walls or placing them in floor screeds to be cemented or plastered in is not permitted. Pipes in these positions need to be protected from the corrosive effects of plaster and cement and they need to have room to expand and contract. The guidance notes give examples on how to deal with pipes in chases or ducts.

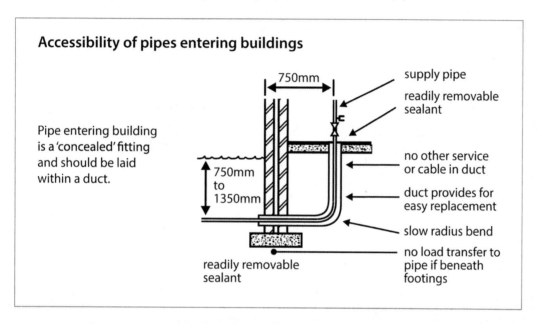

Accessibility of pipes entering buildings

Pipe entering building is a 'concealed' fitting and should be laid within a duct.

750mm

750mm to 1350mm

readily removable sealant

supply pipe

readily removable sealant

no other service or cable in duct

duct provides for easy replacement

slow radius bend

no load transfer to pipe if beneath footings

2. **Valves, joints and other components are required to be** positioned where they are **readily accessible for operation, maintenance and repair.**

3. **Concealed fittings must be made of gunmetal or dezincification resistant material** such as DZR brass.

4. **Water fittings laid below ground must be deep enough to prevent freezing.** (At least 750mm depth of cover)

But! What is a concealed water fitting?

Sub-paragraph 7(5) states that a **'concealed water fitting' is one which:**

(a) **is installed below ground;**

(b) **passes through or under any wall, footing or foundation;**

(c) **is enclosed in any chase or duct; or**

(d) **is in any other position which is inaccessible or renders access difficult**

Accessibility of pipes in or under solid floors

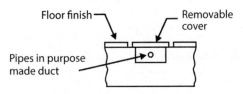

Floor finish

Removable cover

Pipes in purpose made duct

(Thermally insulated if in an unheated building)
Recommended practice

Tilling of other surface finish

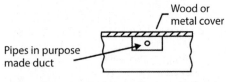

Wood or metal cover

Pipes in purpose made duct

Acceptable only where few joints are enclosed and pipe can be withdrawn for examination

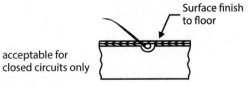

Surface finish to floor

acceptable for closed circuits only

Acceptable where the pipe is wrapped in impervious tape

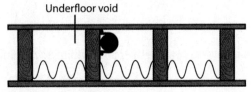

Underfloor void

Pipe thermally insulated in purpose made duct under floor

Acceptable only where few joints are enclosed and pipe can be withdrawn for examination

Accessibility of pipes behind bath panels

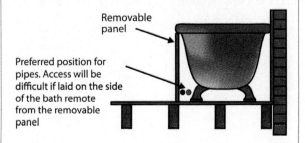

Removable panel

Preferred position for pipes. Access will be difficult if laid on the side of the bath remote from the removable panel

Accessibility of pipes in solid walls

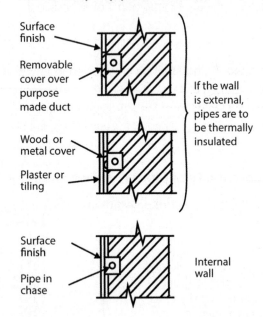

Surface finish

Removable cover over purpose made duct

Wood or metal cover

Plaster or tiling

If the wall is external, pipes are to be thermally insulated

Surface finish

Pipe in chase

Internal wall

NOTE: This is only permitted in internal wall and if pipe can be capped off or isolated should a leak become apparent

Accessibility of pipes in cavity walls

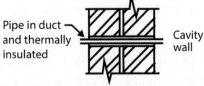

Pipe in duct and thermally insulated

Cavity wall

NOTE: This is only permitted in an internal wall and if pipe can be capped off or isolated should a leak become apparent

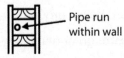

Plasterboard and studding wall (internal)

Pipe run within wall

Accessibility of pipes under suspended floors

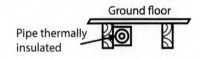

Ground floor

Pipe thermally insulated

Access at intervals of not more than 2m and at every joint for inspection of whole length of pipe

Self-assessment questions

1. Water fittings are required to be 'immune to or protected from galvanic action'.

    a) *Explain briefly what is meant by galvanic action, and* **b)** *give one example*

    a) **Galvanic action is:** _____

    _____

    b) An example which could cause galvanic action is: _____

2. State the minimum pressure that water fittings are required to withstand.

    _____

3. Complete the following statements.

    Water storage cisterns must:

    – *have* _____ _____ *lids and screened* _____ *and* _____

    – *not be placed in positions where they could become* _____

4. Give one good reason why draining valves should not be positioned in sump holes or below ground.

    _____

5. We should avoid installing pipes and fittings in areas that are particularly vulnerable to frost such as unheated parts of the building, drafty areas cold surfaces and any area outside buildings above ground.

    Give TWO examples for each of the general areas listed below:

    – *unheated parts of the building, e.g.*   *i)* _____

                                      ii) _____

    – *known draughty areas within buildings e.g.* *i)* _____

                                      ii) _____

    – *cold surfaces. e.g.*   *i)* _____

                                      ii) _____

6. Depth of pipes below ground.

 a) *Give the minimum and maximum depth of cover for a pipe laid below ground*

  minimum _____ *maximum* _____

 b) *Where the required depths of cover cannot be achieved, suggest what action must be taken*

  _____

7. Pipes rising from below ground outside buildings should be fitted with adequate insulation. State THREE additional requirements that should be taken.

 i) _____

 ii) _____

 iii) _____

8. Complete the following statements.

 a) *Generally, pipes entering buildings should:*

  – *be in a* _____ *that will allow access for renewal and repair, and also give some degree of insulation,*

  – *enter through or under the foundations at a minimum depth of* _____

  – *rise up in the building at least 750mm from the* _____ *, and*

  – *where the pipe rises up through a suspended (ventilated) floor, the pipe should be*

  _____ *from floor level to a depth of* _____

 b) *In roof spaces insulation should be below cisterns and where possible pipes should be kept beneath* _____ *insulation.*

9. Complete the following statements.

 a) *Draining valves should be fitted on all* _____ _____ *so that*

  the _____ *system is drainable.*

 b) *The best protection from frost is to maintain rooms containing water fittings to a* _____ *that frost cannot occur.*

10. Some materials may be permeated and contaminated by contact with oil, petrol or gas. Name the group of materials that are likely to be susceptible to permeation damage giving one example.

The group of materials is _____

an example is _____

11. Pipes and other components should be securely fixed but not so as to restrict thermal movement.

    Indicate with arrows on the diagram which clips are incorrectly positioned and state why.

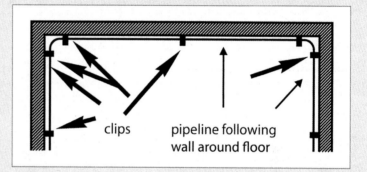

clips   pipeline following wall around floor

The clips I have indicated are incorrectly positioned because

_____

12. Indicate which of the following diagrams shows the correct support for a plastics cistern.

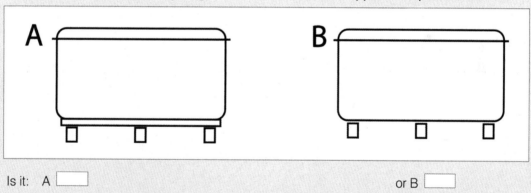

A   B

Is it: A [   ]   or B [   ]

13. When installing pumps or boosters, the water undertaker is likely to insist on a break cistern to separate the incoming supply pipe from a boosted supply. Give TWO reasons for this.

    1. _____

    2. _____

14. The water undertaker must be notified of certain pump or booster installations. Give the conditions under which that notification might apply.

_____

15. Complete the following sentences.

a) No water fitting shall be embedded in any _____ or _____ _____

b) Valves, joints and other components that require to be operated, maintained or repaired should be _____ where they are readily _____ .

c) Concealed water fittings shall be made of _____ or

_____ _____ materials.

16. Complete the following statement from sub-paragraph 7(5) of Schedule 2.

A 'concealed water fitting' is one which:

a) is installed _____ ground;

b) passes _____ or _____ any wall, footing or foundation;

c) is _____ in any chase or duct; or

d) is in any other position which is _____

or renders _____ difficult

Check your answers on pages 27, 28, and 29.

Summary of main points

Paragraphs 3 to 7 of Schedule 2 are concerned with how water fittings are used so that contamination or waste of water will not occur.

Water fittings must be immune to or protected from galvanic action, a term applied to the electrolytic corrosion of dissimilar metals in a damp or wet environment.

Water fittings must resist the effects of external load, vibration, stress, or settlement. Select and install fittings so that stress is kept to a minimum.

Water fittings	– **to be watertight and remain watertight in service**
	– **to be capable of withstanding 1½ times the maximum expected internal pressure**
	– **not to be placed in a position that impurities can pass through the material to contaminate the water**
Water storage cisterns	– **to have rigid, close fitting and securely fixed lids, and be fitted with screened vents and overflows**
	– **not to be placed in positions where they could become submerged**

Cisterns, cylinders and other components should be arranged so they can not become contaminated by build-up of sediment. Regular maintenance is essential.

Draincocks in sump holes beneath floors or below ground level are not permitted because they constitute a contamination risk.

Frost precautions

Avoid installing pipes and fittings in **areas such as**:

– **unheated parts of the building,** e.g. roof spaces, cellars, garages and outhouses;
– **known draughty areas** within buildings e.g. near door, window or ventilation openings, in 'cold' roofs, spaces and beneath suspended ventilated floors.
– **cold surfaces.** e.g. direct contact with outside walls, or ducts or chases within outside walls.
– **any position outside buildings and above ground.**

Pipes below ground should be laid at minimum of 750mm depth of cover, and for access, not more than 1350mm.

Where depths of between 750mm and 1350mm cannot be achieved, the water undertaker must be notified. (Regulation 5)

Pipes rising from below ground outside buildings should be fitted with adequate water proofed insulation that extends to a depth of 750mm and be protected from other damage.

Generally, **pipes entering buildings should**:

– **be in a duct** that will allow access for renewal and repair,
– **enter through or under the foundations** at a minimum **depth of 750mm**
– **rise up** in the building **at least 750mm from the outside wall.**
  (If less than 750mm the pipe should be insulated) and
– **where the pipe rises up through a suspended** (ventilated) **floor,** the pipe should **be insulated** from floor level **to a depth of 750mm** below ground level.

Pipes and fittings should not be chased or ducted into outside walls, and fittings in unheated places should be insulated.

In roof spaces insulation should be omitted below cisterns and where possible pipes kept beneath ceiling insulation.

Pipes above ground outside buildings should be avoided, particularly in draughty places and insulation should be waterproofed.

Insulation should be provided, particularly for pipes and fittings in unheated or partially heated areas or where extreme conditions occur. Table 4.2 of the DEFRA guidance gives advice on insulation thicknesses in normal exposure conditions.

Draining valves should be fitted on all low points so that the whole system is drainable.

The best protection is to maintain rooms containing water fittings to a temperature that frost cannot occur.

Some plastics materials may be damaged by contact with certain other fluids and gases. Where it is known that contamination from oil, petrol (hydrocarbons) or gas could possibly occur, other materials should be used that are resistant to permeation.

Pipes and other components should be securely fixed but not so as to restrict thermal movement. Clips to be spaced as per BS 8558/BS EN 806-4.

Water storage cisterns need proper support, and flexible cisterns should be supported on a continuous platform that supports the whole of the base area.

Boosted water supplies should be fitted with break cisterns to separate the booster pump from the incoming supply pipe.

If a pump or booster supplies more than 12 litres per minute from the supply pipe, the water undertaker must be notified. (Regulation 5)

No water fitting shall be embedded in any wall or solid floor.

Valves, joints and other components that require to be operated, maintained or repaired should be positioned where they are readily accessible.

Concealed water fittings shall be made of gunmetal or dezincification resistant materials.

A 'concealed water fitting' is one which:

(a) is installed below ground;

(b) passes through or under any wall, footing or foundation;

(c) is enclosed in any chase or duct; or

(d) is in any other position which is inaccessible or renders access difficult.

Answers to self-assessment questions

1. Water fittings are required to be 'immune to or protected from galvanic action'.

   a) **Explain briefly what is meant by galvanic action and**   b) give one example

   a) *Galvanic action is* **a term applied to the electrolytic corrosion of dissimilar metals in a damp or wet environment.**

   b) *An example which could cause galvanic action is; (any answer similar to these)*
      - **corrosion due to the use of copper pipes with a galvanized cistern**
      - **copper tube soldered directly into lead pipe**

2. State the minimum pressure that water fittings are required to withstand.

   1½ **times the maximum expected internal pressure**

3. Complete the following statements.

   Water storage cisterns must:
   - *have* **close fitting** *lids and screened* **vents** *and* **overflows**
   - *not be placed in positions where they could become* **submerged**

4. Give one good reason why draining valves should not be positioned in sump holes or below ground.

   Because they could be the cause of contamination

5. We should avoid installing pipes and fittings in areas that are particularly vulnerable to frost such as unheated parts of the building, drafty areas cold surfaces and any area outside buildings above ground:

   Give TWO examples for each of the general areas listed below:
   - unheated parts of the building, e.g. **roof spaces, cellars, garages, outhouses;**
   - known draughty areas within buildings e.g. **near door, window or ventilation openings, in 'cold' roofs, spaces and beneath suspended ventilated floors;**
   - cold surfaces e.g. **direct contact with outside walls, or ducts or chases within outside walls.**

6. Depth of pipes below ground.

   a) Give the minimum and maximum depth of cover for a pipe laid below ground
      minimum of 750mm   maximum 1350mm

   b) Where the required depths of cover cannot be achieved, suggest what action must be taken
      The water undertaker must be notified.

7. Pipes rising from below ground outside buildings should be fitted with adequate insulation. State THREE additional requirements that should be taken.

    i)   *the insulation must extends to a depth of 750mm; and*

    ii)   *the insulation must be waterproofed;*

    iii) *the insulation and waterproofing must be protected from other damage.*

8. Complete the following statements.

    a) *Generally,* **pipes entering buildings should:**
       – *be in a* **duct** *that will allow access and also give some degree of insulation,*
       – *enter through or under the foundations at a minimum* **depth of 750mm**
       – *rise up in the building at least 750mm from* **the outside wall,** *and*
       – *where the pipe rises up through a suspended (ventilated) floor, the pipe should be* **insulated** *from floor level to a depth of* **750mm**

    b) *In roof spaces insulation should be omitted* **below cisterns** *and where possible pipes kept beneath* **ceiling insulation.**

9. Complete the following statements.

    a) *Draining valves should be fitted on all* **low points** *so that the* **whole** *system is drainable.*

    b) *The best protection from frost is to maintain rooms containing water fittings to a* **temperature** *that frost cannot occur.*

10. Some materials may be permeated and contaminated by contact with oil, petrol or gas. Name the group of materials that are likely to be susceptible to permeation damage giving one example.

    The group of materials is   **plastics materials**

    an example is   **PVC-U** *or* **MDPE (blue)**

11. Pipes and other components should be securely fixed but not so as to restrict thermal movement.

    Indicate with arrows on the diagram which clips are incorrectly positioned and state why.

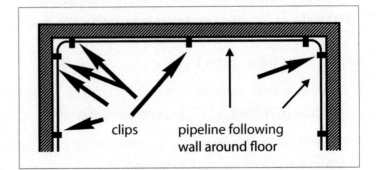

clips   pipeline following wall around floor

The clips I have indicated are incorrectly positioned because **they are too near the bend and will restrict thermal movement.**

12. Indicate which of the following diagrams shows the correct support for a plastics cistern

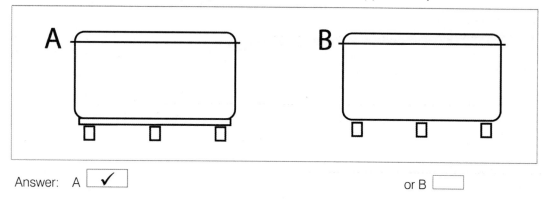

A   B

Answer: A ☑   or B ☐

13. When installing pumps or boosters, the water undertaker is likely to insist on a break cistern to separate the incoming supply pipe from a boosted supply. Give TWO good reasons for this.

 1. *To prevent backflow from the boosted supply pipe.*

 2. *To prevent the pump or booster from causing a vacuum in the main and thus reducing the water pressure in the main or causing backflow from another supply pipe.*

14. The water undertaker must be notified of certain pump or booster installations. Give the conditions under which that notification might apply.

 If a pump or booster supplies more than 12 litres per minute (Regulation 5)

15. Complete the following sentences.

 a) *No water fitting shall be embedded in any* **wall** *or* **solid floor.**

 b) *Valves, joints and other components that require to be operated, maintained or repaired should be* **positioned** *where they are readily* **accessible.**

 c) *Concealed water fittings shall be made of* **gunmetal** *or* **dezincification resistant** *materials.*

16. *Complete the following statement from sub-paragraph 7(5) of Schedule 2.*

 A 'concealed water fitting' is one which:

 a) *is installed* **below** *ground;*

 b) *passes* **through** *or* **under** *any wall, footing or foundation;*

 c) *is* **enclosed** *in any chase or duct; or*

 d) *is in any other position which is* **inaccessible** *or renders* **access** *difficult*

What to do next

There! That's another module completed!

Now to Water system design and installation

Go on to Module 5

Water Industry Act 1991:

Water Supply (Water Fittings) Regulations 1999

An Open Learning Course

Module 5

Water system design and installation

Introduction

Paragraphs 8 to 13 of Schedule 2 deals with a variety of topics under the heading 'water system design and installation'.

Paragraph 8 prescribes that water fittings should be installed so that they cannot become contaminated, or placed in positions where they could become damaged. Damage to pipes and fittings could of course lead to waste of water through leakage.

Paragraph 9 looks at the need to keep cold water from being warmed which is of course a contamination risk.

Paragraph 10 requires the provision of stopvalves on supply pipes and distributing pipes supplying water to separate premises. It also covers the provision of stopvalves for individual supply pipes and in those premises having joint supply pipes.

Paragraph 11 deals with the draining down of systems and requires all pipes and fittings to be fitted with servicing valves and draining valves to minimise waste of water when systems are drained down. It also requires stopvalves to be fitted for the isolation of parts of the pipework.

Finally, testing, flushing and disinfecting of completed or altered water fittings is the subject of **Paragraphs 12 and 13**.

However, because of the importance of this subject, and to help keep the size of individual learning modules to a reasonable size, **Paragraphs 12 and 13** are treated separately in Module 6 under the heading 'Commissioning'.

What is the requirement?

Schedule 2: Materials and substances in contact with water Paragraphs 8 to 11:

8.   *No water fitting shall be installed in such a position, or pass through such surroundings, that it is likely to cause contamination or damage to the material of the fitting or the contamination of water supplied by the water undertaker.*

9.   *Any pipe supplying cold water for domestic purposes to any tap shall be so installed that, so far as is reasonably practicable, the water is not warmed above 25°C.*

10.-(1) *Every supply pipe or distributing pipe providing water to separate premises shall be fitted with a stopvalve conveniently located to enable the supply to those premises to be shut off without shutting off the supply to any other premises.*

(2) *Where a supply pipe or distributing pipe provides water in common to two or more premises, it shall be fitted with a stopvalve to which each occupier of those premises has access.*

11.   *Water supply systems shall be capable of being drained down and be fitted with an adequate number of servicing valves and drain taps so as to minimise the discharge of water when the water fittings are maintained or replaced. A sufficient number of stopvalves shall be installed for isolating parts of the pipework.*

How should systems be designed and installed to avoid contamination and waste of water?

If we look at **Paragraph 8**, we see that **pipes and fittings should be arranged to avoid any position where**:

(a) **the water could become contaminated, or**

(b) **damage to the material could occur**

One of the previous byelaws said words to this effect. 'No supply pipe, distributing pipe or other water fitting shall be laid or installed in or on, or be laid in or through any foul soil, refuse or refuse chute, ash pit, sewer, drain, cesspool, or any manhole connected with any such sewer drain or cesspool.'

This suggests a number of locations that our pipes should not be laid through. The positions listed are quite obviously unhealthy environments which could well lead to contamination of water within a pipe, particularly where the pipe or fitting has become corroded and leakage is taking place.

They are also usually very corrosive environments, and corrosion damage to pipes and fittings can lead to leakage and waste of water.

So! The previous byelaw requirements still provide us with good guidance to the current Regulations.

How else can installations be arranged to avoid damage to pipes and fittings?

There are a great many ways in which our installations can become damaged. A few examples are given below, most of which are everyday examples of the good workmanship we carry out all the time.

– **Lay pipes below ground at sufficient depth (minimum 750mm)** to avoid damage by digging, or by vehicles moving about above a pipeline. Sufficient depth of cover will also guard against frost damage.

– **Pipes passing through walls below ground should be protected from ground movement or settlement** of the building.

Examples of ways of laying pipes below ground through walls or under foundations

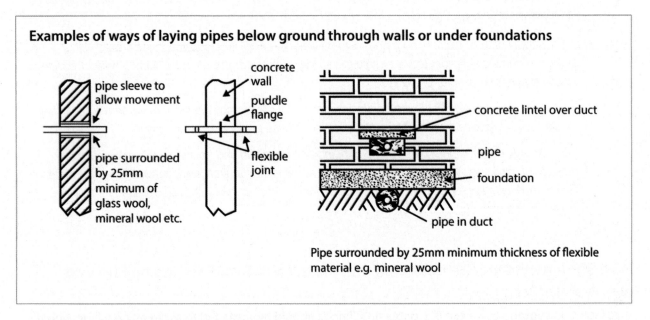

Pipe surrounded by 25mm minimum thickness of flexible material e.g. mineral wool

– **Pipes passing through walls and embedded in walls or floors should be** wrapped or otherwise **protected from any corrosive effects of cement, concrete and dampness** whilst these products dry out following construction works.

– **Pipes in ducts and channels should be arranged so that they can easily be withdrawn** for renewal or repair. This applies particularly to pipes entering buildings.

– **Avoid contact between plastics pipes and oil or petrol** which can soften and weaken the pipe material leading to leakage and contamination risk. Gases from oil, petrol and gas installations can permeate (penetrate) plastic pipe without any obvious damage, but leave a taste or smell to contaminate the water.

- **Pipes fittings and appliances should be adequately supported** without restricting thermal movement but at the same time preventing undue vibration that might lead to noise or damage. Pipes (particularly below ground) should be anchored at bends, tees, valves and pipe ends to prevent them pushing apart under pressure.

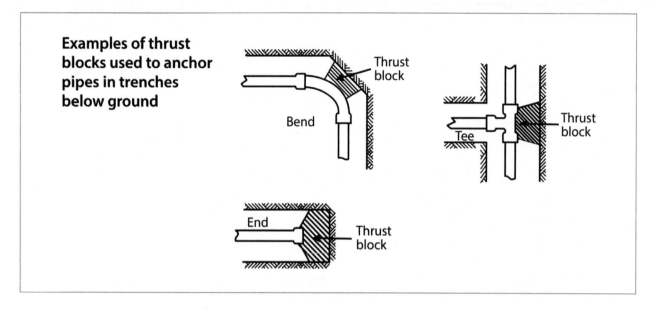

Examples of thrust blocks used to anchor pipes in trenches below ground

Thrust block

Bend

Thrust block

Tee

End

Thrust block

- **Thermal movement** is a problem that cannot be prevented particularly in plastic pipes or in systems that are continually heated and cooled. **Pipework should be arranged so that expansion and contraction can freely take place** without damage to the pipe or its fixings.

- **Pipes should be arranged to avoid airlocks** allowing air to escape when filling and air to enter when emptying systems. Pipes should rise and fall towards vents and outlets.

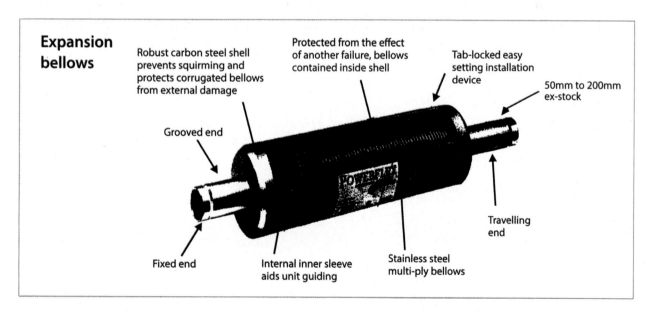

Expansion bellows

Robust carbon steel shell prevents squirming and protects corrugated bellows from external damage

Protected from the effect of another failure, bellows contained inside shell

Tab-locked easy setting installation device

50mm to 200mm ex-stock

Grooved end

Fixed end

Internal inner sleeve aids unit guiding

Stainless steel multi-ply bellows

Travelling end

The examples are endless, but by using common sense and good installation practice it should be quite easy to make sure your installations comply with this requirement.

Cold pipes need to be kept below 25°C, but how?

Paragraph 9 deals with this and is quite straightforward and easy to understand. It says *'any pipe supplying cold water for domestic purposes shall be so installed that, so far as is reasonably practicable, the water is not warmed above 25°C.'*

Easy to understand, yes, but unfortunately not always so easy to apply. Why? because in practice we have no control over the weather, nor over the heating of the building once the installation is put into use. But, **there are precautions we can take**, and as long as we take them, we shall have met our obligation under this requirement.

So **to prevent pipes and fittings from becoming unduly warmed**, we can:

– **position cold pipes below hot ones** to avoid rising heat;

– **route pipes runs around heated areas** rather than through them;

– **insulate pipes** and fittings to reduce the effects of heat gains, particularly those in roof spaces that can get extremely hot in summer;

– advise the customer of the importance of maintaining the system to keep it safe from contamination.

Note: The Guidance Document says the temperature of cold water should be kept below 20°C **as does BS 8558:2011. However, that is good practice, the regulations are law.**

But why do we need to keep cold pipes cool?

The answer is simple. A warm pipe will promote the growth of bacterium in water much more quickly than a cold ones. Critical temperatures are between 25°C and 50°C. Between these temperatures bacterium such as legionella will flourish and can create unhealthy situations which should be avoided.

It should also be borne in mind that legionella bacterium can lie dormant at temperatures well below 25°C and remain in the water just waiting to be warmed so they can flourish and become a danger! Hence the need to keep the temperature of cold water to a minimum.

Which pipes are we talking about?

Any pipes that carry cold water and other fittings and appliances as well. In particular, pipes passing through airing cupboards and heated rooms. There is also a special need to insulate pipes and cisterns in roof spaces and in airing cupboards.

Remember! It is just as important to insulate pipes against heat gains as it is to protect them from frost.

Self-assessment questions (1)

1. Paragraph 8 gives give two positions which must be avoided when arranging pipes and fittings. Complete the following statements.

   Pipes and fittings should be arranged to avoid any position where:

   a) _____

   b) _____

2. The previous byelaws were quite specific about positions to avoid when laying pipes and fittings. Give TWO positions that should be avoided.

   a) _____

   b) _____

3. The following precautions can be taken to avoid contamination and/or waste of water. Indicate in the spaces provided whether they are likely to cause contamination or waste or both.

	contamination	waste
Provide adequate depth of pipes below ground		
Avoid contact between plastic pipes and oil or petrol		
Avoid contact between plastic pipes and gas		
Wrap embedded pipes to protect from cements etc.		
Arrange pipes so thermal expansion can take place		
Provide proper support for pipes, fittings and appliances		

4. Paragraph 9 states that cold water should not be warmed. State:

   a) what maximum temperature is given for cold water, _____

   b) why the water should be kept cool?

   _____

5. Give TWO parts of a building where cold pipes are particularly vulnerable to heat and where precautions should be taken to prevent the water being warmed.

   1._____ 2._____

Check your answers on page 26.

Stopvalves, and servicing valves

What is the difference between a stopvalve and a servicing valve and what is important about fitting them?

The definitions in Paragraph 1 describe both the stopvalve and servicing valve and these are repeated below. There are certain positions in which stopvalves and servicing valves must be fitted to properly serve their purpose and there are certain types of valve that can be used as a stopvalve and others that can be used as servicing valves.

They are defined in Paragraph 1 as follows:

"servicing valve" means a valve for shutting off, for the purpose of maintenance or service, the flow of water in a pipe connected to a water fitting.

Please note it says 'for maintenance or service' which means it is there purely as a means of shutting off the supply for those purposes.

"stopvalve" means a valve, other than a servicing valve, used for shutting off the flow of water in a pipe. Generally the Regulations deal with those stopvalves that are required to be fitted to control the whole supply of water to premises, and those that are required to isolate separate sections of an installation within premises.

Let's look first at stopvalves used to control the supply to individual premises

Paragraph 10.-(1) says that *'Every supply pipe or distributing pipe providing water to separate premises shall be fitted with a stopvalve conveniently located to enable the supply to those premises to be shut off without shutting off the supply to any other premises'.*

Paragraph 10.-(2) says *'Where a supply pipe or distributing pipe provides water in common to two or more premises, it shall be fitted with a stopvalve to which each occupier of those premises has access'.*

It will **normally** be the case that **separate premises**, whether they be a house or some other type of building or property, **will be supplied with water through a single and separate supply pipe** that is adequate in size to suit the needs of that building and that building alone.

Some buildings however, because of their circumstances, may be permitted to be supplied from a common supply pipe. The onus is on the water supplier to make a decision whether a common pipe might be permitted after looking at the individual situation. In the vast majority of cases a separate supply will be required.

Whichever the case, **every** house, flat, factory, shop, office or other **premises**, that is occupied separately, **must have a stopvalve fitted that will shut off the supply to the whole of the premises.**

It is important also that occupiers of premises have access to a stopvalve that they can turn off the supply in the event of leakage, or for other reasons, without causing any disturbance or nuisance to any other premises. **The stopvalve must not shut off the supply to any other premises.**

Where should the stopvalve be fitted?

A stopvalve should be provided **on the supply pipe**, and be positioned:

- **inside the premises;**
- **above floor level;**
- **as near as possible to the point of entry to the premises;**
- **to control the supply to the whole premises; and**
- **should NOT shut off the supply to any other premises.**

The following diagrams illustrate how and where stopvalves should be provided.

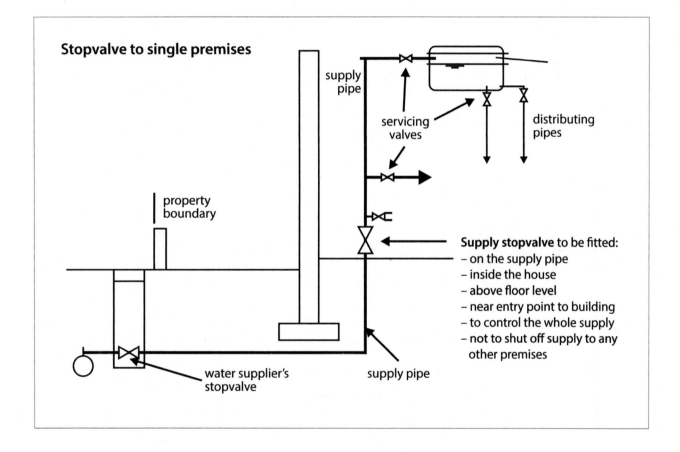

Stopvalve to single premises

supply pipe

servicing valves

distributing pipes

property boundary

Supply stopvalve to be fitted:
– on the supply pipe
– inside the house
– above floor level
– near entry point to building
– to control the whole supply
– not to shut off supply to any other premises

water supplier's stopvalve

supply pipe

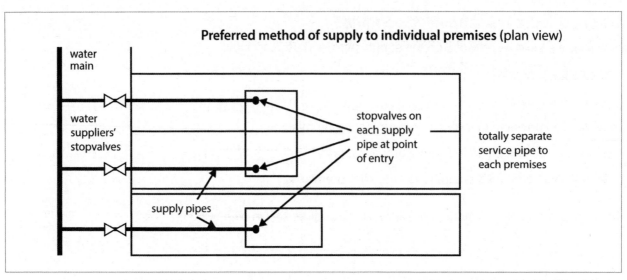

Preferred method of supply to individual premises (plan view)

- water main
- water suppliers' stopvalves
- supply pipes
- stopvalves on each supply pipe at point of entry
- totally separate service pipe to each premises

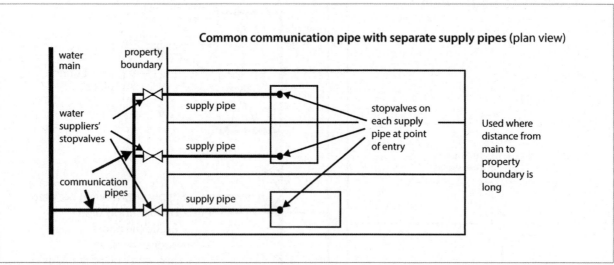

Common communication pipe with separate supply pipes (plan view)

- water main
- property boundary
- water suppliers' stopvalves
- communication pipes
- supply pipe
- supply pipe
- supply pipe
- stopvalves on each supply pipe at point of entry
- Used where distance from main to property boundary is long

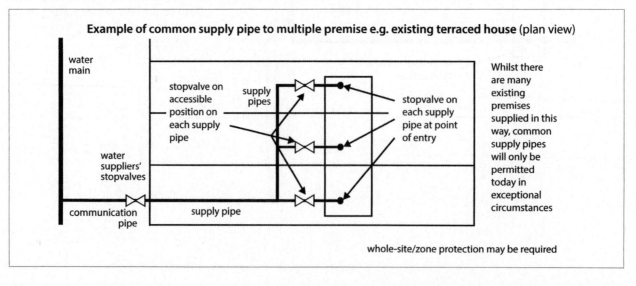

Example of common supply pipe to multiple premise e.g. existing terraced house (plan view)

- water main
- stopvalve on accessible position on each supply pipe
- supply pipes
- water suppliers' stopvalves
- communication pipe
- supply pipe
- stopvalve on each supply pipe at point of entry
- Whilst there are many existing premises supplied in this way, common supply pipes will only be permitted today in exceptional circumstances

whole-site/zone protection may be required

Note: The above diagrams refer to service pipes, supply pipes and communication pipes. These three terms were all referred to and defined in previous water byelaws but the new Regulations apply only to the supply pipe, and not the communication pipe which is the responsibility of the water undertaker. For a fuller description of the service pipe, supply pipe and communication pipe you should look in the Glossary of Terms attached to this package.

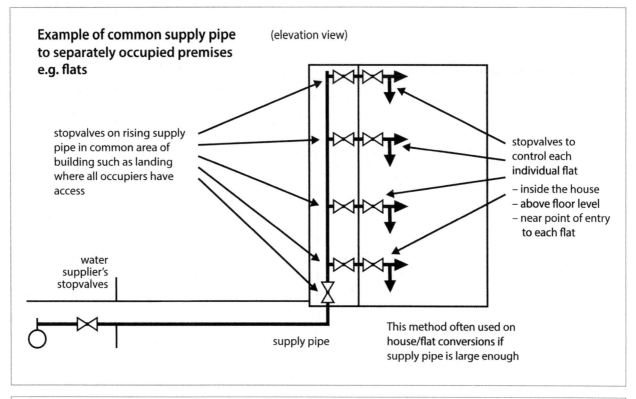

Example of common supply pipe to separately occupied premises e.g. flats

(elevation view)

stopvalves on rising supply pipe in common area of building such as landing where all occupiers have access

stopvalves to control each individual flat

– inside the house
– above floor level
– near point of entry to each flat

water supplier's stopvalves

supply pipe

This method often used on house/flat conversions if supply pipe is large enough

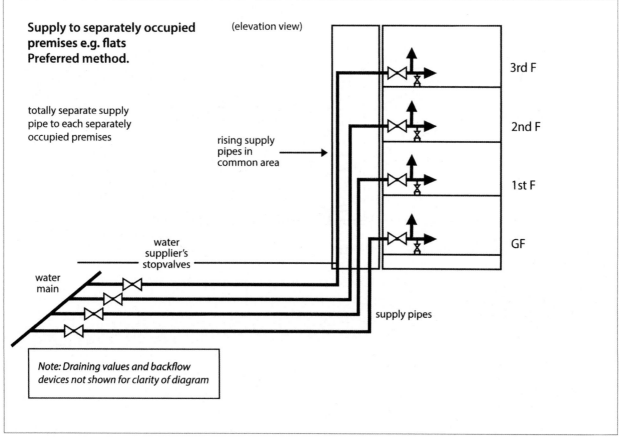

Supply to separately occupied premises e.g. flats Preferred method.

(elevation view)

totally separate supply pipe to each separately occupied premises

rising supply pipes in common area

3rd F

2nd F

1st F

GF

water supplier's stopvalves

water main

supply pipes

Note: Draining values and backflow devices not shown for clarity of diagram

Where distributing pipes supply separately chargeable premises from a common storage cistern each separate premises are required to be fitted with stopvalves in similar positions to that described for stopvalves on common supply pipes. These will usually be tall buildings that have fittings above the limit of the mains supply.

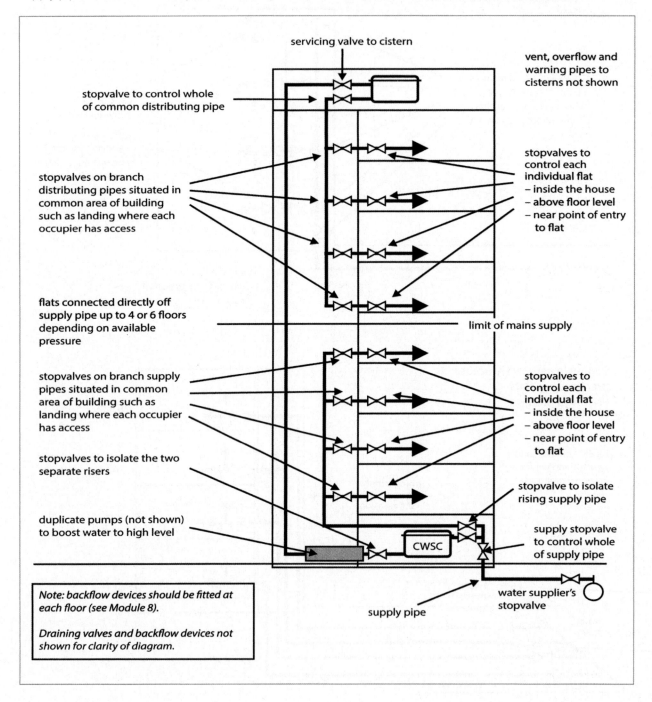

servicing valve to cistern

vent, overflow and warning pipes to cisterns not shown

stopvalve to control whole of common distributing pipe

stopvalves to control each individual flat
– inside the house
– above floor level
– near point of entry to flat

stopvalves on branch distributing pipes situated in common area of building such as landing where each occupier has access

flats connected directly off supply pipe up to 4 or 6 floors depending on available pressure

limit of mains supply

stopvalves on branch supply pipes situated in common area of building such as landing where each occupier has access

stopvalves to control each individual flat
– inside the house
– above floor level
– near point of entry to flat

stopvalves to isolate the two separate risers

stopvalve to isolate rising supply pipe

duplicate pumps (not shown) to boost water to high level

CWSC

supply stopvalve to control whole of supply pipe

Note: backflow devices should be fitted at each floor (see Module 8).

Draining valves and backflow devices not shown for clarity of diagram.

supply pipe

water supplier's stopvalve

Supplies to separate buildings within single premises

Some premises comprise of two or more buildings or have a number of different parts that need to be supplied with water. Although these separate buildings or parts are not separately chargeable they to will need to have stopvalves fitted to isolate sections of the pipework and thus comply with Paragraph 11.

Practices used for separately-chargeable premises should be followed for those premises containing parts that are not separately chargeable.

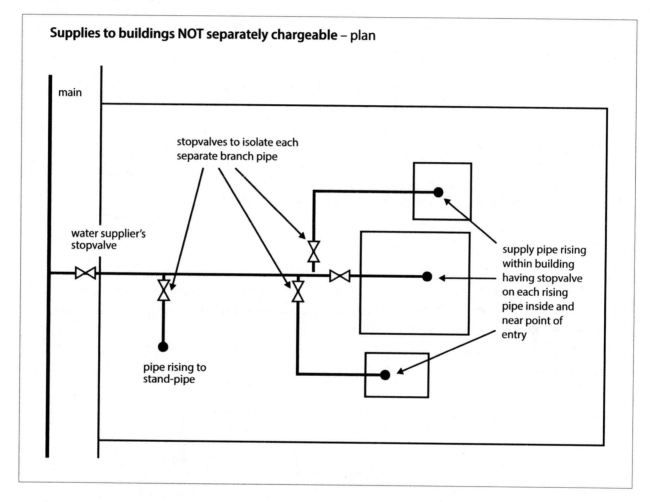

Supplies to buildings NOT separately chargeable – plan

main

stopvalves to isolate each separate branch pipe

water supplier's stopvalve

supply pipe rising within building having stopvalve on each rising pipe inside and near point of entry

pipe rising to stand-pipe

Self-assessment questions (2)

1. There is one position where a stopvalve is required to be fitted to the supply pipe within premises. Complete the following statement to give the five rules that apply to the positioning of that stopvalve on the supply pipe.

   A stopvalve should be provided on the supply pipe, and be positioned:

   _____

   _____

   _____

   _____

   _____

2. In some buildings water is supplied to separate premises from a storage cistern. Will the same five rules apply to distributing pipes in these cases as applies to the supply pipe?

   Answer Yes or No   *Yes* ☐   *No* ☐

3. Assume you have to lay a supply pipe to each of four new properties, say TWO shops with TWO flats above the shops. Using the diagram below, describe, by adding pipeline drawings, how these premises should be supplied. Show the method of supply most likely to be accepted by the water undertaker, and indicate the position of stopvalves.

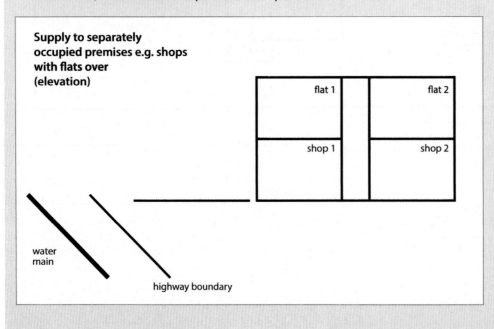

Supply to separately occupied premises e.g. shops with flats over (elevation)

flat 1   flat 2

shop 1   shop 2

water main

highway boundary

4. The next diagram shows a block of flats (separately occupied) which is partly supplied directly from the supply pipe and partly from a high level cold water storage cistern.

   Add all necessary stopvalves to comply with Paragraph 10(2).

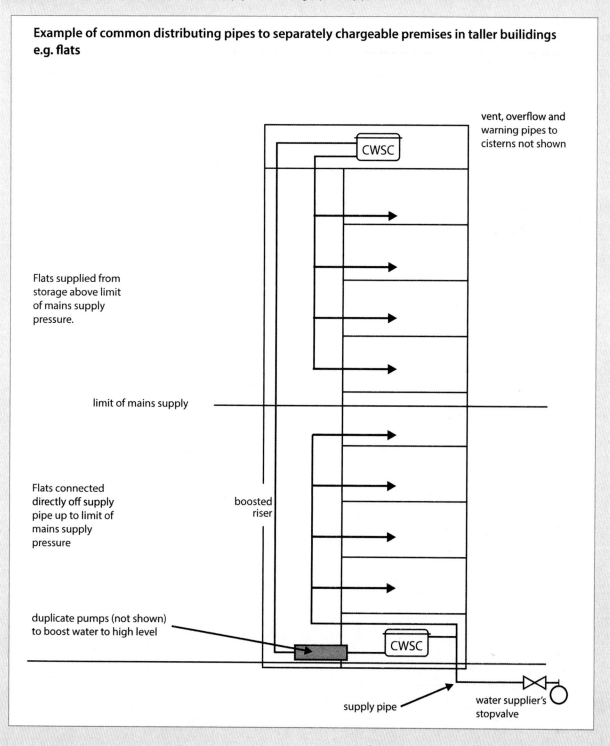

Example of common distributing pipes to separately chargeable premises in taller builidings e.g. flats

vent, overflow and warning pipes to cisterns not shown

CWSC

Flats supplied from storage above limit of mains supply pressure.

limit of mains supply

Flats connected directly off supply pipe up to limit of mains supply pressure

boosted riser

duplicate pumps (not shown) to boost water to high level

CWSC

supply pipe

water supplier's stopvalve

Check your answers on pages 27 and 28.

What about servicing valves and draining valves?

Paragraph 11 is concerned with the provision of stopvalves and servicing valves, and states that **servicing valves and draining valves should be fitted** in adequate numbers **to minimise waste of water when draining down systems** or parts of a system **for maintenance and repair**.

Suitable positions for servicing valves and draining valves will give full control over all or part of the installation, and **allow sections of pipework** or individual appliances **to be turned off and drained down completely** without disturbing the supply to other parts of the premises.

It should be remembered at this point that draining valves are also used as a frost precaution.
(See Module 3 'Materials')

There are a number of positions where the fitting of servicing valves is especially important.

Servicing valves should be fitted:

– **immediately before every float operated valve;**

– **on every cold feed pipe and distributing pipe from any water storage cistern (except on cold feed pipes to primary heating circuits);**

– **on every hot water distributing pipe where it is otherwise impossible to fit a valve on a cold feed pipe. e.g. on combination unit.**

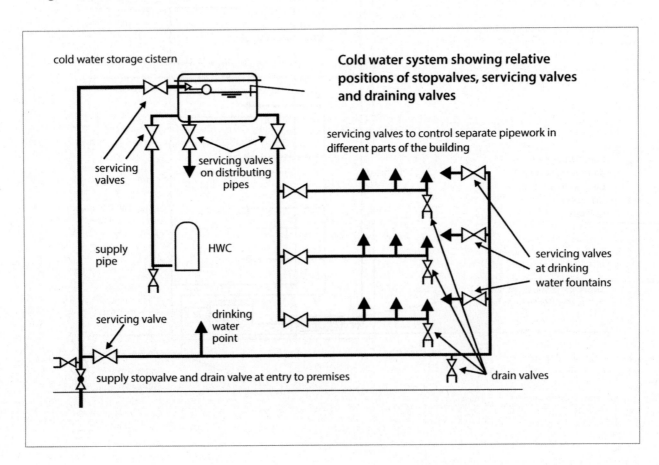

cold water storage cistern

Cold water system showing relative positions of stopvalves, servicing valves and draining valves

servicing valves to control separate pipework in different parts of the building

servicing valves

servicing valves on distributing pipes

servicing valves at drinking water fountains

supply pipe

HWC

servicing valve

drinking water point

supply stopvalve and drain valve at entry to premises

drain valves

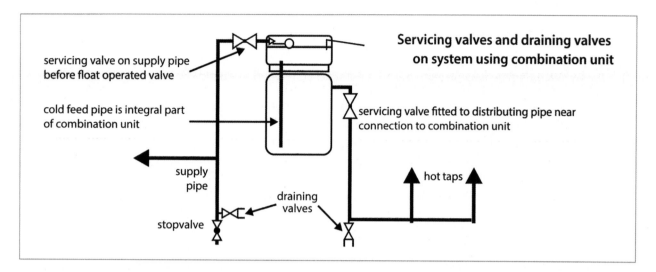

servicing valve on supply pipe before float operated valve

cold feed pipe is integral part of combination unit

Servicing valves and draining valves on system using combination unit

servicing valve fitted to distributing pipe near connection to combination unit

supply pipe

hot taps

draining valves

stopvalve

Valves to isolate parts of the system

Additionally, **Paragraph 11** states that **stopvalves must be installed to isolate parts of the pipework,** for maintenance and for closing down sections of the supply when leakages are taking place.

On larger installations, servicing valves or stopvalves should be fitted:

– to isolate pipework on different floors;

– to isolate various parts of an installation at the same level;

– to isolate branch pipes to ranges of appliance.

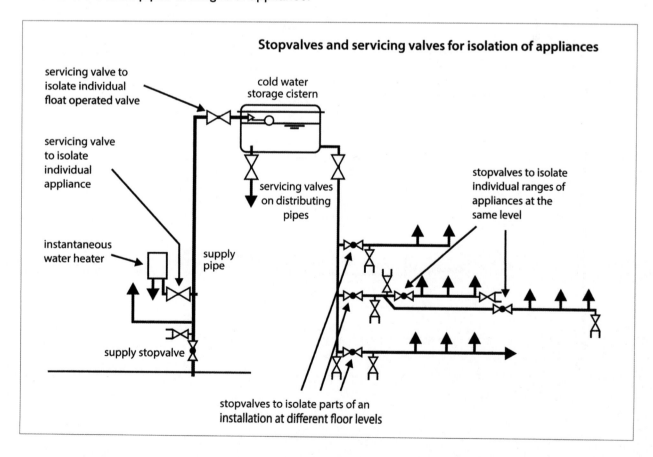

Stopvalves and servicing valves for isolation of appliances

servicing valve to isolate individual float operated valve

cold water storage cistern

servicing valve to isolate individual appliance

stopvalves to isolate individual ranges of appliances at the same level

servicing valves on distributing pipes

instantaneous water heater

supply pipe

supply stopvalve

stopvalves to isolate parts of an installation at different floor levels

Self-assessment questions (3)

1. Paragraph 11 states that systems must be capable of being drained down.

    State why servicing valves and draining valves are required to be fitted in adequate numbers for maintenance and repair.

    _____

2. Paragraph 11 also mentions stopvalves. What does it say about them?

    _____

3. Give THREE positions where servicing valves should be fitted for maintenance and servicing purposes.

    a) _____

    b) _____

    _____

    c) _____

4. Add stopvalves, servicing valves and draining valves in suitable positions on the following cold water system diagram.

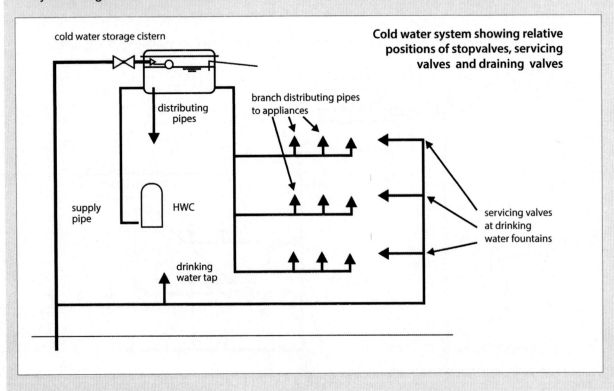

cold water storage cistern

Cold water system showing relative positions of stopvalves, servicing valves and draining valves

distributing pipes

branch distributing pipes to appliances

supply pipe

HWC

servicing valves at drinking water fountains

drinking water tap

5. Add stopvalves, servicing valves and draining valves in suitable positions on the following diagram of a combination unit.

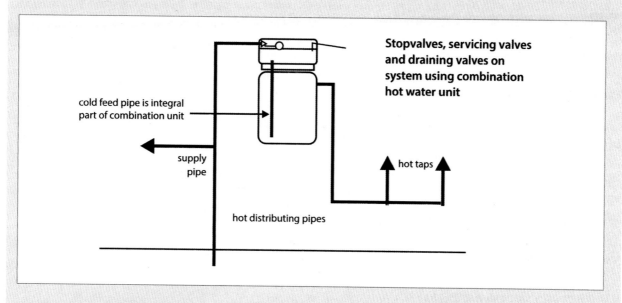

cold feed pipe is integral part of combination unit

Stopvalves, servicing valves and draining valves on system using combination hot water unit

supply pipe

hot taps

hot distributing pipes

6. On larger installations stopvalves are required to isolate parts of the installation. Indicate the position of all necessary stopvalves, servicing valves and draining valves on the following diagram.

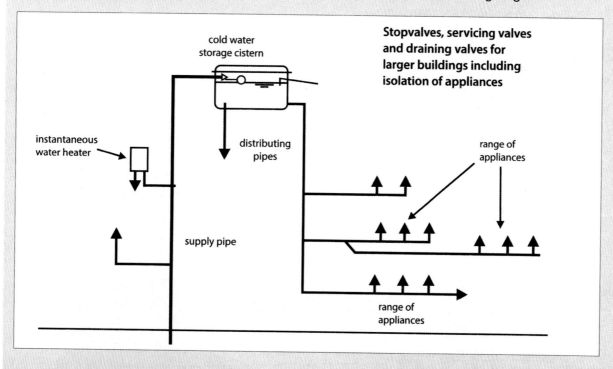

cold water storage cistern

Stopvalves, servicing valves and draining valves for larger buildings including isolation of appliances

instantaneous water heater

distributing pipes

range of appliances

supply pipe

range of appliances

Check your answers on pages 29 and 30.

What valves are suitable for use as stopvalves or servicing valves?

There are a number of valves that are considered suitable for use as stopvalves and servicing valves. Each of these should comply with a relevant British or EN Standard and care should be taken in selection of a valve for a particular situation.

Any valve used below ground or in inaccessible positions should be made of gunmetal or of a material that is resistant to dezincification. Spherical plug valves used to isolate WC cisterns should be of the slot type so as to inhibit unauthorised consumer interference.

Stopvalves below ground should be of a type that will permit shutting off by use of a key (crutch type or square cap type).

Above 50mm diameter
In any position, above or below ground

1. **flanged gate valve to BS 5163**, predominantly key operated

Up to 50mm diameter
In any position above or below ground

2. **screwdown stopvalve to BS 5433**
3. **plugcock to BS 2580; or**

Above ground only

4. **screwdown stopvalve to BS 1010 Part 2** or
5. **wheel operated (gate) valve to BS 5154**
6. **screwdriver-operated, slot type spherical plug valve to BS 6675**
7. **lever-operated, spherical plug valve to BS 6675**

A selection of stopvalves and servicing valves are illustrated below.

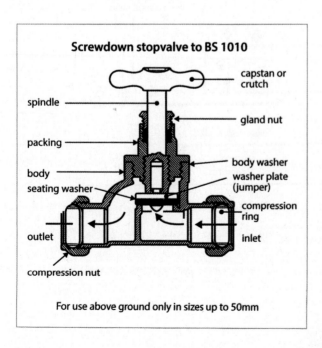

Screwdown stopvalve to BS 1010

- capstan or crutch
- spindle
- gland nut
- packing
- body washer
- body
- washer plate (jumper)
- seating washer
- compression ring
- outlet
- inlet
- compression nut

For use above ground only in sizes up to 50mm

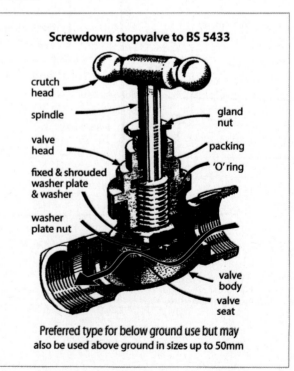

Screwdown stopvalve to BS 5433

- crutch head
- gland nut
- spindle
- valve head
- packing
- fixed & shrouded washer plate & washer
- 'O' ring
- washer plate nut
- valve body
- valve seat

Preferred type for below ground use but may also be used above ground in sizes up to 50mm

Flanged gate valve to BS 5163

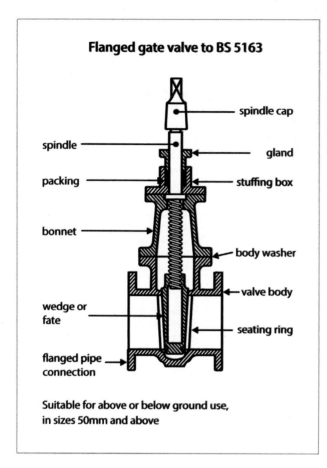

- spindle cap
- spindle
- gland
- packing
- stuffing box
- bonnet
- body washer
- valve body
- wedge or fate
- seating ring
- flanged pipe connection

Suitable for above or below ground use, in sizes 50mm and above

Plugcock to BS 2580

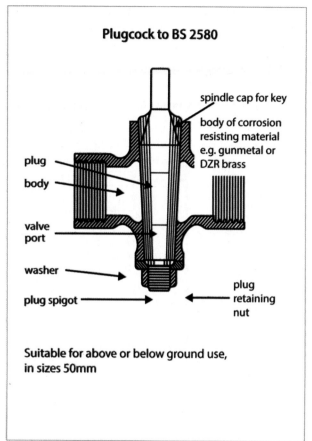

- spindle cap for key
- body of corrosion resisting material e.g. gunmetal or DZR brass
- plug
- body
- valve port
- washer
- plug spigot
- plug retaining nut

Suitable for above or below ground use, in sizes 50mm

Screwdriver-operated, slot type spherical plug valve to BS 6675

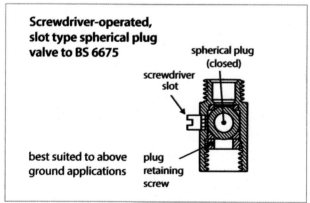

- spherical plug (closed)
- screwdriver slot
- plug retaining screw

best suited to above ground applications

Wheel operated (gate) valve to BS 5154

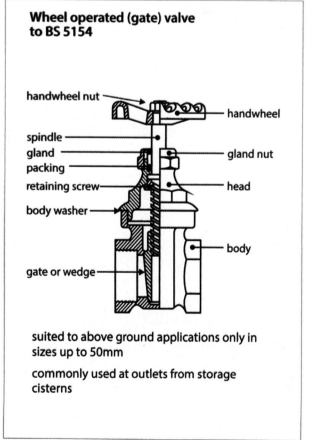

- handwheel nut
- handwheel
- spindle
- gland
- gland nut
- packing
- retaining screw
- head
- body washer
- body
- gate or wedge

suited to above ground applications only in sizes up to 50mm

commonly used at outlets from storage cisterns

Lever-operated spherical plug valve to BS 6675

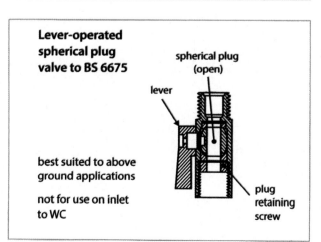

- spherical plug (open)
- lever
- plug retaining screw

best suited to above ground applications

not for use on inlet to WC

To sum up the question of stopvalves and servicing valves, let's look at another diagram and a chart showing the uses and positioning of valves. The chart suggests what the Regulations will accept and the diagram gives a more traditional look at where valves should be positioned based on past experience and good practice.

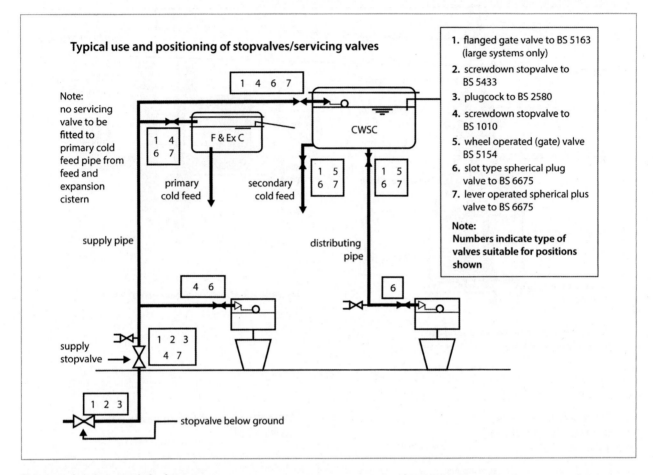

Typical use and positioning of stopvalves/servicing valves

Note: no servicing valve to be fitted to primary cold feed pipe from feed and expansion cistern

F & Ex C

CWSC

primary cold feed

secondary cold feed

supply pipe

distributing pipe

supply stopvalve

stopvalve below ground

1. flanged gate valve to BS 5163 (large systems only)
2. screwdown stopvalve to BS 5433
3. plugcock to BS 2580
4. screwdown stopvalve to BS 1010
5. wheel operated (gate) valve BS 5154
6. slot type spherical plug valve to BS 6675
7. lever operated spherical plus valve to BS 6675

Note: Numbers indicate type of valves suitable for positions shown

Types of valve and their uses

	Type of valve	Above ground	Below ground	Above 50mm	Below 50mm	Supply stopvalve	Servicing valve	Isolating valve
1	flanged gate valve to BS 5163	yes	yes	yes	no	yes	yes	yes
2	screwdown stopvalve to BS 5163	yes	yes	no	yes	yes	yes	yes
3	plugcock to BS 2580	yes	yes	no*	yes	yes	yes	yes
4	screwdown stopvalve to BS 1010	yes	no	no	yes	yes	yes	yes
5	wheel operated valve (gate valve) to BS 5154	yes	no	no	yes	no	yes	yes
6	spherical plug valve (slot type) to BS 6675	yes	no	no	yes	no	yes	yes
7	spherical plug valve (lever operated) to BS 6675	yes	no	no	yes	no	yes	yes

Self-assessment questions (4)

1. There are a number of types of stopvalve and servicing valve. Name ONE suitable valve for each of the positions indicated in the diagram below.

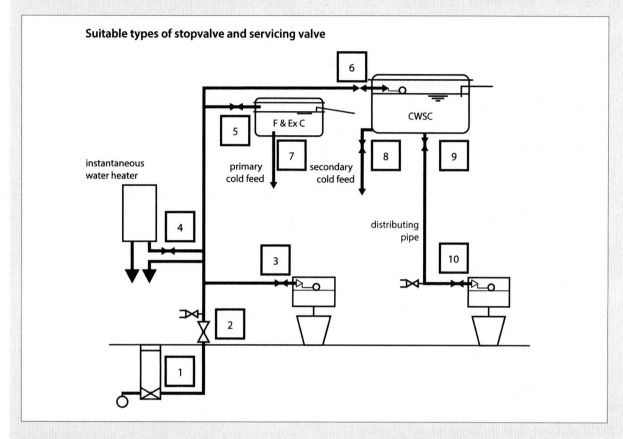

Suitable types of stopvalve and servicing valve

Answers:

1. _____

2. _____

3. _____

4. _____

5. _____

6. _____

7. _____

8. _____

9. _____

10._____

Check your answers on page 31.

Summary of main points

Paragraph 8 prescribes that **pipes and fittings should be arranged to avoid any position where:**

(a) the water could become contaminated, or

(b) damage to the material could occur.

When installing pipes and fittings the following points should be born in mind:

- **lay pipes below ground at sufficient depth (minimum 750mm).**
- **pipes passing through walls below ground should be protected from ground movement or settlement.**
- **pipes passing and embedded in walls or floors should be protected from corrosive effects of cement, concrete and damp.**
- **avoid contact between plastic pipes and oil or petrol.**
- **pipes, fittings and appliances should be adequately supported** without restricting thermal movement.
- **pipework should be arranged so that expansion and contraction can freely take place.**
- **pipes should be arrange to avoid air locks.**

Paragraph 9. **Any pipe supplying cold water for domestic purposes shall be installed so the water is not warmed above 25°C.**

To prevent pipes and fittings from becoming unduly warmed:

- **position cold pipes below hot ones** to avoid rising heat.
- **route pipes runs around heated areas** rather than through them.
- **insulate pipes** and fittings to reduce the effects of heat gains.

Why? Because a warm pipe will promote the growth of bacterium in water much more quickly than cold ones. **Critical temperatures are between 25°C and 50°C.**

Paragraph 10 requires the provision of stopvalves on supply pipes and distributing pipes supplying water to separate premises.

A stopvalve should be provided on the supply pipe, and be positioned:

- **inside the premises;**
- **above floor level;**
- **as near as possible to the point of entry to the premises;**
- **to control the supply to the whole premises; and**
- **should NOT shut off the supply to any other premises.**

Normally **separate premises will be supplied with water through a single, separate supply pipe** that is adequate in size to suit the needs of the building. Only in exceptional circumstances will a common pipe be permitted.

Every house, flat, factory, shop, office or other **premises**, that is occupied separately, **must have a stopvalve fitted that will shut off the supply to the whole of the premises.**

The stopvalve must not shut off the supply to any other premises.

In some cases, particularly in flats, premises may be supplied from storage in the roof. **In these cases the distributing pipe to each of the premises must also be fitted with a stopvalve to control the whole of the supply to that premises.**

In Paragraph 11, **servicing valves and draining valves should be fitted** in adequate numbers to **minimise waste of water when draining down**.

Suitable positions for servicing valves and draining valves will allow sections of pipework or individual appliances to be turned off and drained down completely for repair and maintenance without disturbing the supply to other parts of the premises.

Additionally, **servicing valves must be installed to isolate parts of the pipework,** for maintenance and for closing down sections of the supply when leakages are taking place.

Servicing valves should be fitted:
- **immediately before every float operated valve;**
- **on every cold feed pipe and distributing pipe from any water storage cistern (except on cold feed pipes to primary heating circuits);**
- **on every hot water distributing pipe where it is otherwise impossible to fit a valve on a cold feed pipe,** e.g. on combination unit.

On larger installations stopvalves should be fitted:
- **to isolate pipework on different floors;**
- **to isolate various parts of an installation at the same level;**
- **to isolate branch pipes to ranges of appliance;**
- **to isolate individual fittings for regular maintenance.**

Stopvalves and servicing valves are generally one of the following types and may be used as seen below.

Types of valve and their uses

	Type of valve	Above ground	Below ground	Above 50mm	Below 50mm	Supply stopvalve	Servicing valve	Isolating valve
1	flanged gate valve to BS 5163	yes	yes	yes	no	yes	yes	yes
2	screwdown stopvalve to BS 5163	yes	yes	no	yes	yes	yes	yes
3	plugcock to BS 2580	yes	yes	no	yes	yes	yes	yes
4	screwdown stopvalve to BS 1010	yes	no	no	yes	yes	yes	yes
5	wheel operated valve (gate valve) to BS 5154	yes	no	no	yes	no	yes	yes
6	spherical plug valve (slot type) to BS 6675	yes	no	no	yes	no	yes	yes
7	spherical plug valve (lever operated) to BS 6675	yes	no	no	yes	no	yes	yes

Answers to self-assessment questions (1)

1. Requirement 8 gives give two positions which must be avoided when arranging pipes and fittings. Complete the following statements.

   Pipes and fittings should be arranged to avoid any position where:

   a) **water could become contaminated**

   b) **damage to the material could occur**

2. The previous byelaws were quite specific about positions to avoid when laying pipes and fittings. Give TWO positions that should be avoided.

   Answer to include any two of; **foul soil, refuse or refuse chute, ash pit, sewer, drain, cesspool, or manhole connected to a drain or sewer.**

3. The following precautions can be taken to avoid contamination and/or waste of water. Indicate in the spaces provided whether they are likely to cause contamination or waste or both.

	contamination	waste
Provide adequate depth of pipes below ground		yes
Avoid contact between plastic pipes and oil or petrol	yes	yes
Avoid contact between plastic pipes and gas	yes	
Wrap embedded pipes to protect from cements etc.		yes
Arrange pipes so thermal expansion can take place		yes
Provide proper support for pipes, fittings and appliances		yes

4. Requirement 9 states that cold water should not be warmed. State:

   a) what maximum temperature is given for cold water   **25°C**

   b) why the water should be kept cool?   **to prevent growth of bacterium**

5. Give TWO parts of a building where cold pipes are particularly vulnerable to heat and where precautions should be taken to prevent the water being warmed.

   1. **roof spaces**

   2. **airing cupboards or similar reply**

Answers to self-assessment questions (2)

1. There is one position where a stopvalve is required to be fitted to the supply pipe within premises. Complete the following statement to give the five rules that apply to the positioning of that stopvalve on the supply pipe.

   A stopvalve should be provided on the supply pipe, and be positioned:

   - *inside the premises;*
   - *above floor level;*
   - *as near as possible to the point of entry to the premises;*
   - *to control the supply to the whole premises; and*
   - *should NOT shut off the supply to any other premises.*

2. In some buildings water is supplied to separate premises from a storage cistern. Will the same five rules apply to distributing pipes in these cases as applies to the supply pipe. Answer Yes or No.

   Yes!

3. Assume you have to lay a supply pipe to each of four new properties, say TWO shops with TWO flats above the shops. Using the diagram below, describe, by adding pipeline drawings how these premises should be supplied. Show the method of supply most likely to be accepted by the water undertaker, and indicate the position of stopvalves.

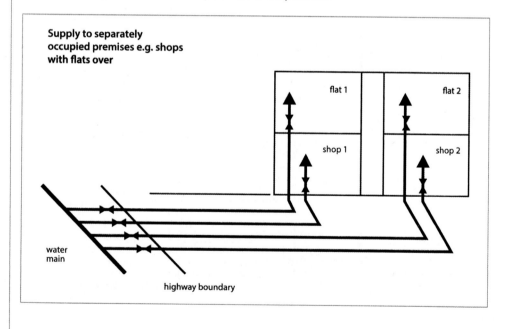

4. The next diagram shows a block of flats (separately occupied) which is partly supplied directly from the supply pipe and party from a high level cold water storage cistern. Add all necessary stopvalves to comply with Paragraph 10(2).

Example of common distributing pipes to separately premises in taller buildings e.g. flats

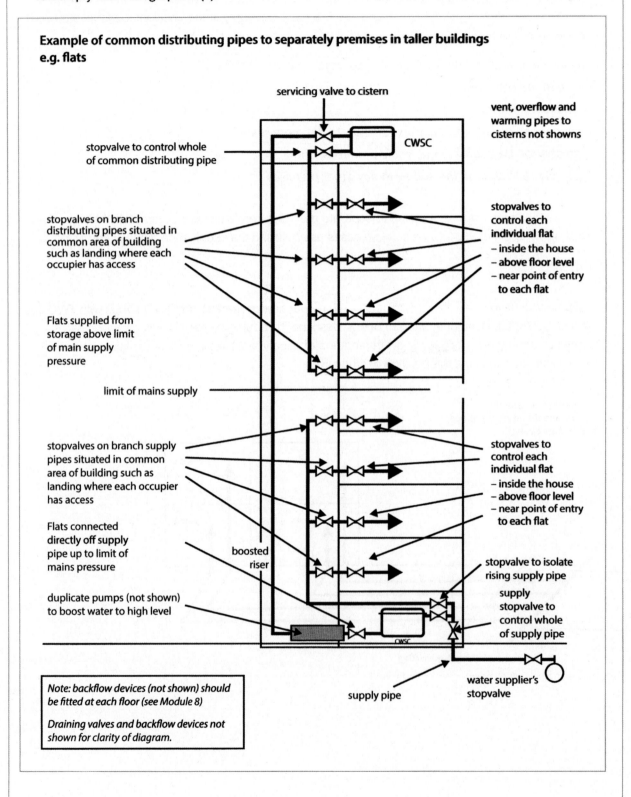

servicing valve to cistern

vent, overflow and warming pipes to cisterns not showns

CWSC

stopvalve to control whole of common distributing pipe

stopvalves on branch distributing pipes situated in common area of building such as landing where each occupier has access

stopvalves to control each individual flat
– inside the house
– above floor level
– near point of entry to each flat

Flats supplied from storage above limit of main supply pressure

limit of mains supply

stopvalves on branch supply pipes situated in common area of building such as landing where each occupier has access

stopvalves to control each individual flat
– inside the house
– above floor level
– near point of entry to each flat

Flats connected directly off supply pipe up to limit of mains pressure

boosted riser

stopvalve to isolate rising supply pipe

supply stopvalve to control whole of supply pipe

duplicate pumps (not shown) to boost water to high level

CWSC

water supplier's stopvalve

supply pipe

Note: backflow devices (not shown) should be fitted at each floor (see Module 8)

Draining valves and backflow devices not shown for clarity of diagram.

Answers to self-assessment questions (3)

1. Paragraph 11 states that systems must be capable of being drained down.

   State why servicing valves and draining valves are required to be fitted in adequate numbers for maintenance and repair.

   To minimise waste of water when draining down.

2. Requirement 11 also mentions stopvalves. What does it say about them?

   Sufficient number of stopvalves must be fitted to isolate parts of the building

3. Give THREE positions where servicing valves should be fitted for maintenance and servicing purposes.

   a) *immediately before every float operated valve;*

   b) *on every cold feed pipe and distributing pipe from any water storage cistern (except on cold feed pipes to primary heating circuits);*

   c) *on every hot water distributing pipe where it is otherwise impossible to fit a valve on a cold feed pipe. e.g. on combination unit.*

4. Add stopvalves, servicing valves and draining valves in suitable positions on the following cold water system diagram.

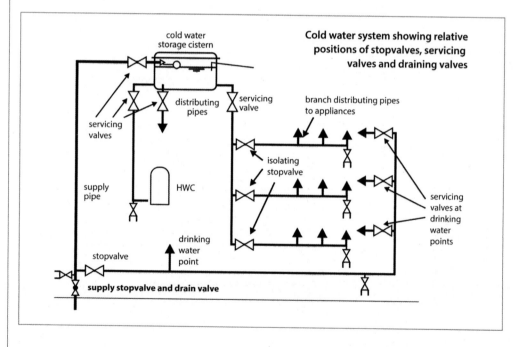

Cold water system showing relative positions of stopvalves, servicing valves and draining valves

5. Add stopvalves, servicing valves and draining valves in suitable positions on the following diagram of a combination unit.

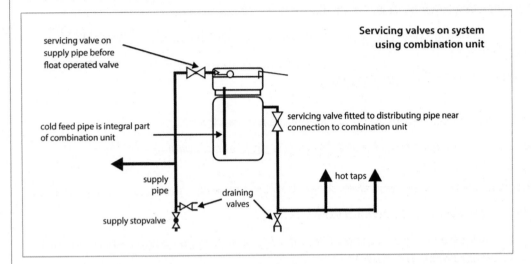

Servicing valves on system using combination unit

servicing valve on supply pipe before float operated valve

cold feed pipe is integral part of combination unit

servicing valve fitted to distributing pipe near connection to combination unit

supply pipe

draining valves

hot taps

supply stopvalve

6. On larger installations, servicing valves and/or stopvalves are required to isolate parts of the installation.

Indicate the position of all necessary stopvalves, servicing valves and draining valves on the following diagram.

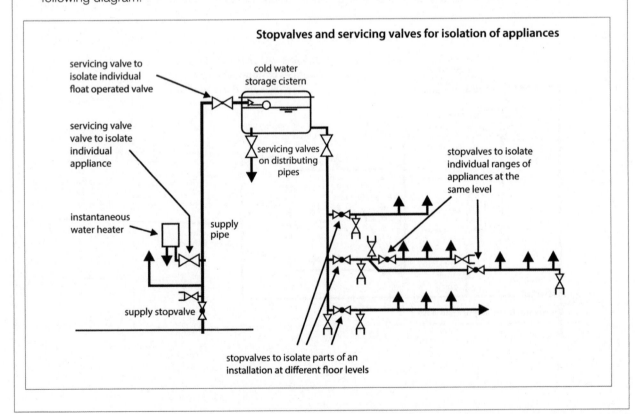

Stopvalves and servicing valves for isolation of appliances

servicing valve to isolate individual float operated valve

cold water storage cistern

servicing valve valve to isolate individual appliance

servicing valves on distributing pipes

stopvalves to isolate individual ranges of appliances at the same level

instantaneous water heater

supply pipe

supply stopvalve

stopvalves to isolate parts of an installation at different floor levels

Answers to self-assessment questions (4)

1. There are a number of types of servicing valve. Name ONE suitable type for each of the positions indicated in the diagram below.

Suitable types of stopvalve and servicing valve

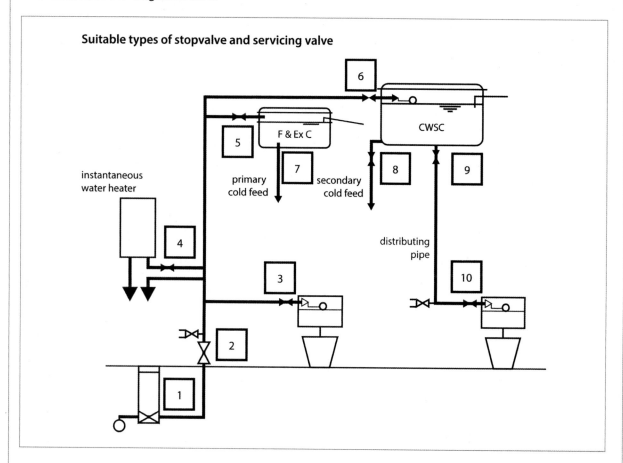

Answers to include any suitable valve. The list given below simply shows one suitable valve for each situation

1. stopvalve to BS 5433

2. stopvalve to BS 1010

3. slot type spherical plug valve to BS 6675

4. slot type spherical plug valve to BS 6675

5. lever operated spherical plug valve to BS 6675

6. lever operated spherical plug valve to BS 6675

7. none

8. wheel operated (gate) valve to BS 5154

9. wheel operated (gate) valve to BS 5154

10. slot type spherical plug valve to BS 6675

What to do next

Good! That's Module 5 under your belt!

Now go on to look at the final part of **Water system design and installation** in Module 6 Commissioning

Water Industry Act 1991:

Water Supply (Water Fittings) Regulations 1999

An Open Learning Course

Module 6

Commissioning

Introduction

Commissioning means finalising an installation, checking it for faults, and putting the system into use.

It means making sure the installation is to the customer's satisfaction, and that it operates safely and efficiently.

Commissioning of hot and cold water installations includes the following processes:

- Visual inspection
- **Soundness testing**
- **Flushing and disinfection**
- Performance testing
- Final checks and handing over

All of these process are considered to be good practice and should be considered for all water installations no matter how large or small the installation is.

Water Regulations are concerned with those processes that might affect the quality of water or waste of water, namely **soundness testing, flushing and disinfection**.

Despite the fact that it is good practice, commissioning is not always carried out properly, and sometimes not done at all. Water Regulations follow the old water byelaws in attempting to encourage every installer to carry out essential commissioning processes.

The testing of water installations has been a requirement of Water Byelaws for many years, and pipes used for the supply of water were required to be flushed 'to remove debris before use after installation, renewal or repair'.

The new Water Regulations, in Paragraph 13 of Schedule 2, goes a stage further by requiring *water fittings to be tested, flushed, and where necessary disinfected before use.*

Paragraph 12 looks specifically at soundness testing.

So! On with the show!

What is the requirement?

Schedule 2: Materials and substances in contact with water Paragraphs 12 to 13:

12.-(1) *The water system shall be capable of withstanding an internal water pressure of not less than 1½ times the maximum pressure to which the installation or relevant part is designed to be subjected in operation. ("The test pressure")*

(2) *This requirement shall be deemed to be satisfied:*

    (a) *in the case of a water system that does not include a pipe made of plastics, where:*

        (i) *the whole system is subjected to the pressure test by pumping, after which the test continues for one hour without further pumping;*

        (ii) *the pressure in the system is maintained for one hour; and*

        (iii) *there is no visible sign of leakage*

    (b) *in any other case, where either of the following tests is satisfied:*

TEST A	TEST B
i) the whole system is subjected to the test pressure by pumping for 30 minutes, after which the test continues for 90 minutes without further pumping,	i) the whole system is subjected to the test pressure by pumping for 30 minutes, after which the pressure is noted and the test continues for 150 minutes without further pumping,
ii) the pressure is reduced to one third of the test pressure after 30 minutes,	ii) the drop in pressure is less than 0.6 bar (60kPa) after the following 30 minutes, or
iii) the pressure does not drop below one one-third of the test pressure over the following 90 minutes; and	iii) 0.8 bar (80kPa) after the following 150 minutes; and
iv) there is no visible leakage throughout the test	iv) **(iv) t**here is no visible leakage throughout the test

13. *Every water system shall be tested, flushed and where necessary disinfected before it is first used*

What does this requirement mean to the practising installer?

You will see above, that criteria for the pressure testing of water systems is set out in Paragraph 12 with some further explanation in the Guidance document.

For full guidance on testing, flushing and disinfection methods one should perhaps refer to BS 8558/BS EN 806-4. It should be remembered however, that **British Standards specifications only give guidance**, they themselves are not law.

However, if you carry out testing, flushing and disinfection in a workmanlike manner [Regulation 4(5)] **and follow the procedures recommended in BS 8558/BS EN 806-4, you are doing what can reasonably be expected to satisfy the law**.

Note: Manufacturer's advice and recommendations regarding chemicals used and the flushing/disinfecting of appliances/equipment should also be considered.

What do I need to know about testing installations for soundness?

Paragraph 12(1) says **that the water system shall be capable of withstanding an internal pressure of 1.5 times the maximum operating pressure ("the test pressure")**.

The test pressure of 1.5 times the maximum operating pressure **applies to ALL tests on ALL installations**.

It does not make any distinction between large or small installations, nor between above or below ground, or new or replacement work. The inference is therefore that ALL installation work should be tested to the test pressure of 1.5 times the maximum operating pressure.

Paragraph 12(2) sets out test criteria and gives separate criteria for:

(a) systems that contain no plastics, e.g. copper, steel, cast iron etc, and

(b) those that do contain pipes or fittings of plastics.

Systems containing no plastics

For non plastics systems, Paragraph 12(2)(a) sets out three test requirements:

1. **Installations shall be pumped up to a test pressure of 1.5 times the maximum operating pressure,** or the maximum operating pressure plus an allowance for any expected surge pressure, whichever is the greatest.

2. for a test period of one hour,

3. **during which, there should be – no visible leakage,** and
                                     – **no loss of pressure**.

Remember! Testing should be carried out on all completed installations including supply pipes, distributing pipes, fittings, components and connections to appliances.

These are **minimum requirements** that should be carried out **to comply with the law**. It does not mean that these are the only things you should do, but whichever system of testing you adopt, the law will expect the five points listed to be included.

How can I carry out a test that will be acceptable under the law?

A procedure based on recommendations of BS 8558/BS EN 806-4, and including the requirement set out above is described on the following page.

Follow this and you won't go far wrong!

Soundness Testing – a procedure based on BS EN 806-4

Before testing you should check that:

(a) All jointing is complete with pipes and components properly secured.

(b) Arrange for the system to be vented.

(c) Pipes and components are visually inspected prior to testing.

(d) Valves within the installation are fully open to ensure the whole section is tested.

Test procedure A: Hydrostatic pressure testing of metal piping systems

(e) Fill the whole installation with water ensuring that all air is removed.

(f) Seal all vents and outlet valves.

(g) Apply the selected test pressure TP equal to 1.1 times the maximum design pressure MDP by pumping for a period of 10 minutes.

(h) Leave for 10 minutes where test pressure must stay constant.

(i) Check for visible leakage and loss of pressure.

(j) If there is a pressure loss, the system shall be maintained at the test pressure until the obvious leaks within the system are identified.

After testing:

(k) a careful record should be kept of both visual and tests carried out.

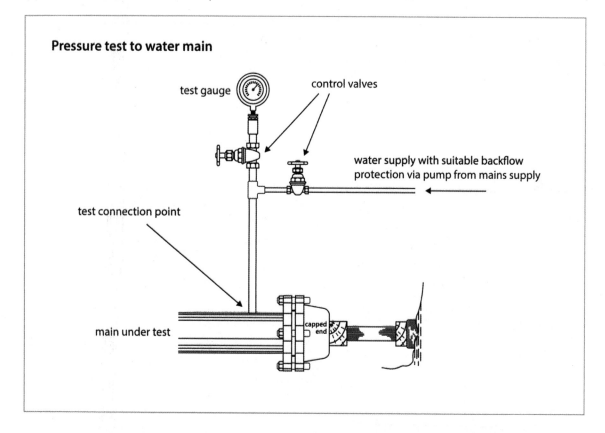

Pressure test to supply pipe

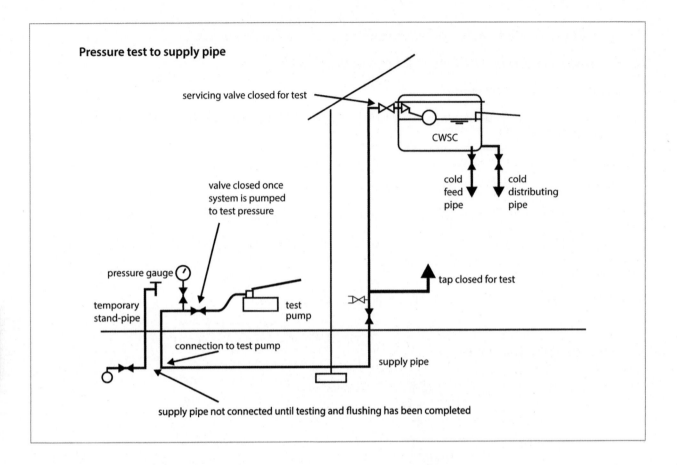

Pressure test to hot water distributing system

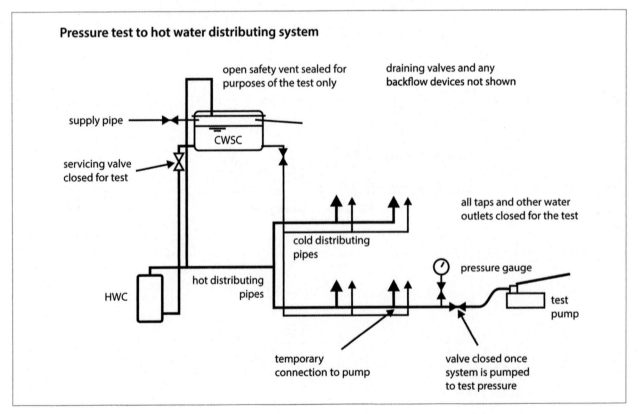

What about the testing of systems containing plastics

Paragraph 12(2)(b) gives criteria for two alternative tests for systems containing plastics and guidance notes suggest that this paragraph will be satisfied if systems containing plastics pipes are tested in accordance with the recommendations of BS EN 806-4.

BS EN 806-4 uses the term '**elastic or visco-elastic**', or as Paragraph 12 refers to them, **plastics pipes**, two common examples being those made of PVC-U and **PVC-C** and **Polyethylene**.

Why then do BS EN 806-4 and the Water Byelaws give different procedures for the testing of elastic or visco-elastic or plastic pipes?

Well! Some plastics pipe materials suffer from stresses when subjected to test pressures, stresses that may be retained in the pipe material after the test is over. These stresses can lead to pipe failure in the future. For this reason pipes of 'elastomeric' materials are permitted to be subjected to test that is a little less severe and that will leave the material with less stress problems after the test.

BS EN 806-4 sets out two alternative test methods for elastic/visco-elastic pipes.

Test procedure B: Hydrostatic pressure testing of plastic piping systems, testing for water soundness

(a) Fill the system with water, ensuring that all air is removed and seal all air vents and outlet valves

(b) Apply the selected test pressure TP equal to 1.1 times the maximum design pressure MDP by pumping, for a period of 30 minutes

(c) An inspection should be carried out to identify any obvious leaks within the system under test

(d) Reduce the pressure by bleeding water from the system to 0.5 times test pressure

(e) Close the bleed valve. The system will be regarded as leak-tight if the pressure maintains a value equal to or greater than 0.5 times the operating pressure for a period of 30 minutes after the pressure reduction. Check visually for leaks. If during that period there is a pressure drop, there will be a leak within the system. Maintain the pressure and identify the leak.

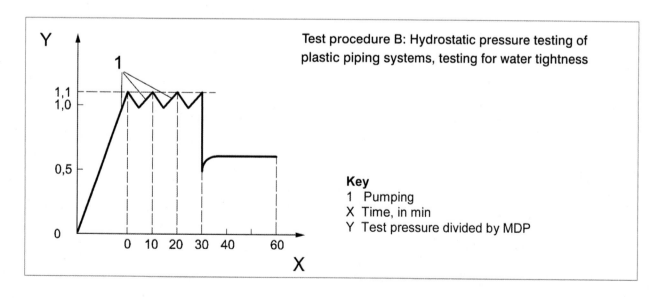

Test procedure B: Hydrostatic pressure testing of plastic piping systems, testing for water tightness

Key
1 Pumping
X Time, in min
Y Test pressure divided by MDP

Test procedure C: Hydrostatic pressure testing of plastic piping systems, testing for water tightness

(a) Fill the system with water, ensuring that all air is removed and seal all air vents and outlet valves.

(b) Apply the selected test pressure TP equal to 1.1 times the maximum design pressure MDP by pumping, for a period of 30 minutes.

(c) Note the pressure after a period of 30 minutes. An inspection should be carried out to identify any obvious leaks within the system.

(d) Note the pressure after a further 30 minutes. If the pressure drop is less than 0.06 MPa (0.6 bar), the system can be considered to have no obvious leaks. Continue the test without further pumping.

(e) Check visually for leaks during the next 2 hours. If the pressure drops by more than 0.02 MPa (0.2 bar) over that period, this will indicate a leak within the system. Maintain the pressure and identify the leak.

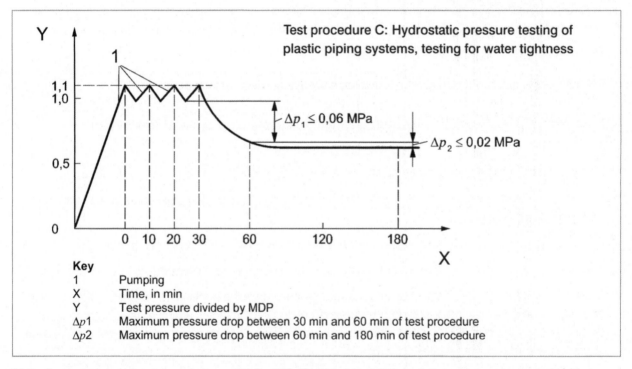

Key

1	Pumping
X	Time, in min
Y	Test pressure divided by MDP
$\Delta p1$	Maximum pressure drop between 30 min and 60 min of test procedure
$\Delta p2$	Maximum pressure drop between 60 min and 180 min of test procedure

Note: It may help readers to read clause 6.1.3 of BS EN 806-4 for full descriptions of recommended testing procedures, and those comprising both plastic and metal pipes.

Are Water Regulations interested in ALL the pressure tests I do?

The short answer is no! But that really needs more explanation, doesn't it. There are two types of test, interim tests and final tests.

Interim tests are applied to pipelines and installations as work progresses and should be done before any work is covered up. These are used to confirm that your work is sound and to help show the client or his representative that the work is progressing satisfactorily.

Final tests are applied when the job is complete and should generally be carried out immediately before the hand-over date. In the case of buried pipelines this means after backfilling, compacting and surface finishes.

Paragraphs 12 and 13 are concerned with making sure that the installation is sound, not just for the test, but for the many years of use after testing has been completed and forgotten. It is therefore the final test that is important under the Regulations.

What is important about flushing and disinfection?

Flushing

Paragraph 13 of Schedule 2 states that *'every water system shall be..., flushed out, ..before it is first used'.*

This means **every new installation, alteration, extension, or any maintenance to existing installations**, is required to be **thoroughly flushed** to remove any debris that might cause contamination of water.

Disinfecting

Paragraph 13 says **every water system shall be... where necessary disinfected before it is first used.**

The Guidance Document suggests that after flushing, systems should be disinfected in:

(a) **new installations**
(except private dwellings occupied by a single family)

(b) **major extensions or alterations**
(except private dwellings occupied by a single family)

(c) **underground pipework**
(except localised repairs or insertion of junctions,) [although there may be a case for larger diameter pipes]

(d) **where contamination may be suspected**
e.g. – fouling by sewage, drainage, animals, or
– after physical entry by personnel for interior inspection, painting or repairs

(e) **where a system has not been in regular use and not regularly flushed**
(regular use means periods of up to 30 days without use depending on the characteristics of the water)

Note: Private dwelling would include the normal house containing a single family. Where a larger house is broken up into separate units, disinfection would be required, as would a house being occupied by other than a single family e.g. students.

For disinfection methods the Guidance Document refers to BS 8558, which in turn means that **if the flushing and disinfection procedures set out in BS 8558 are followed then the Regulations are deemed to be satisfied**.

So go to the next page to see what BS 8558 has to say!

BS 8558   Recommendations for disinfection

Firstly BS 8558 gives the following guidance on sterilisation of pipes and fittings.

– Where pipework is under mains pressure or has a backflow device fitted downstream, the water undertaker should be notified.

– If water used for disinfection is to be discharged to a sewer or drain or water course the authority responsible for the sewer, drain or water course should be notified.

– Chemicals used for disinfection of drinking water installations must be chosen from a list of substances compiled by the Drinking Water Inspectorate and which is listed in the Water Fittings and Materials Directory published by the Water Regulations Advisory Scheme. (WRAS)

In practice, unless any other chemical is specified, a chemical disinfectant such as sodium hyperchlorite (diluted chlorine) may be used. Alternatively, disinfectant in tablet form is probably the best bet. They are relatively easy to obtain and reasonably safe to use.

The sequence of disinfection should follow the water flow into the premises.

1. water mains
2. service pipes and cisterns
3. distribution system

For single dwellings and minor alterations and repairs, disinfection will not be necessary unless contamination is suspected. However the system should be flushed out!

Are there any safety points to watch out for?

- **Yes!** Systems or parts of systems should be taken out of use during any disinfection period, and all outlets marked DISINFECTION IN PROGRESS – DO NOT USE.
- Personnel carrying out the disinfection process must receive appropriate health and safety training under the COSHH Regulations.
- No other chemicals (e.g. toilet cleansers) should be added to the water during disinfection as this may cause generation of toxic fumes.
- ALL users of the premises should be notified that disinfection is to take place.
- Care should be taken when using disinfection materials which can be hazardous to the health and safety of the user. Use safety goggles and protective clothing and refrain from smoking during the disinfection process.

Disinfection procedure (BS 8558:2011)

Procedure using chlorine as a disinfectant

This method can be used for all hot and cold water installations.

1. Thoroughly flush system prior to disinfection (see page 9).
2. Fill system systematically with chlorinated water to initial concentration of 50mg per litre (50ppm) ensuring that the whole system is saturated.
3. Leave for one hour. If the free residual chlorine measured at the end of the contact period is less than 30mg/l then the disinfection procedure should be repeated.
4. Immediately following successful disinfection, thoroughly flush with clean water, continuing to flush until the residual chlorine level is at the same level as the drinking water supplied.

After flushing, samples should be taken for bacteriological analysis, and where analysis provides unsatisfactory results the disinfection and sampling procedures should be repeated.

For supply pipes including any hot or cold storage vessels directly connected and under mains pressure, the chlorine solution should be injected into the lower end of the pipe near its point of connection to the communication pipe using a properly installed injector point. The above procedure 1 to 4 can then be followed.

Water should be fed into the system by systematically opening taps and working away from the point of connection until the **whole** of the distribution system is filled with water of the specified concentration. (50ppm)

For cistern fed gravity systems within buildings, the cistern should be filled with chlorinated water to the concentrations shown above and water fed into the system by systematically opening taps and working away from the cistern until the **whole** of the system is filled.

The above procedure 1 to 4 can then be followed.

The cisterns water level should be maintained using chlorinated water throughout the filling process.

Chlorination can be corrosive to both pipes, fittings and any internal coatings that might have been used within the water system. Any coatings should be properly cured before any disinfection processes are commenced.

Note: It may help readers to read clause 5.2.3.3 of BS 8558:2011 for full descriptions of recommended disinfection procedures, which considers methods using both chlorine and other disinfectants.

Chlorine and other disinfectants can be hazardous so consider the safety points outlined on page 9.

Disinfection process

Disinfecting supply pipe

1. Thoroughly flush and empty at least twice

2. Fill the system with clean water. Cistern to be full to overflowing level.

3. Close all draw-off taps, and valves.

4. Inject chlorine at rate of 50ppm.

5. Open draw-offs in turn starting at the bottom and check for residual chlorine at each draw-off point.

6. Leave for contact period of one hour.

7. After contact period, check residual chlorine at each draw-off point. If less than 30ppm the disinfection process should be repeated.

8. Drain and flush system until residual chlorine is at same level as incoming drinking water.

9. Take sample for bacteriological analysis.

Temporary connection from mains supply with backflow prevention device

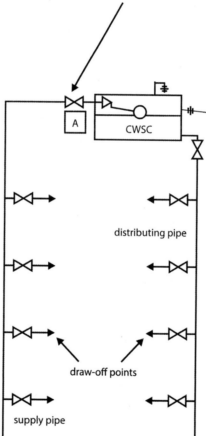

Servicing valve to remain closed except when filling or topping up

A   CWSC

distributing pipe

draw-off points

supply pipe

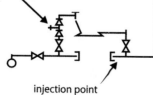

injection point

Disinfecting distributing pipe

1. Thoroughly flush and empty at least twice.

2. Fill the system with clean water. Cistern to be full to overflowing level.

3. Close all draw-off taps, valves, and shut servicing valve (A).

4. Add chlorine to cistern at rate of 50ppm.

5. Open draw-offs in turn starting at the top (nearest the cistern) and check for residual chlorine at each draw-off point.

Note: Cistern may need to be topped up as work proceeds.

6. Leave for contact period of one hour.

7. After contact period, check residual chlorine at each draw-off point. If less than 30ppm the disinfection process should be repeated.

8. Drain and flush system until residual chlorine is at same level as incoming drinking water.

9. Take sample for bacteriological analysis.

If direct connection from main is to be made, backflow protection should be provided and the water supplier should be notified.

The water supplier may require an interposed cistern to be used.

Water supply hygiene

Providing public water supplies that are safe to drink requires a constant vigilance to assure the wholesomeness of supplies.

Within the United Kingdom, which does have a very good record of providing safe drinking water there is an ever present risk of waterborne illness. The development of water supply hygiene safeguards have guided water suppliers (water undertakers) in the practices and procedures they should follow.

Within the United Kingdom there is a comprehensive regulatory system for maintaining water supply quality based around the proactive management of risk and a holistic approach to drinking water safety. With maintenance of robust hygiene practices which forms an essential control measure in the protection of human health.

Robust water hygiene practices underpin many aspects of a risk-based approach to managing water supplies and these principles are therefore intended to support water undertakers in the development and maintenance of drinking water safety plans. Water undertakers must continue to be vigilant and recognise and assess all hazards (natural or anthropogenic) that may lead to the contamination of water supplies.

Therefore it is essential that all those involved in the production and distribution of drinking water, whether employees of a water undertaker or of a contractor, should understand the need to undertake precautions.

Hygienic practices
- **Working practices**
  - Minimum, the number of personnel that are required to work on both water and sewage installations
  - Working on both water and sewage installations personnel to be properly trained and authorised

- **Clothing**
  - Under no circumstances should clothing that has been used on wastewater operations be worn within defined restricted areas
  - Clothing should be clean, free from contamination and be stored apart from other protective clothing
  - Protective clothing (including footwear) used by staff on restricted operations should be readily identifiable

- **Toilets and washing**
  - Water undertakers and contractors should ensure adequate toilet/washing facilities
  - Where portable or temporary arrangements are in place, wastes must be disposed of without risk to water supplies or to the environment

- **Tools and equipment**
  - Stored in designated areas for restricted operations
  - Used in restricted areas should be similarly clean, free from contamination and disinfected before use
  - Separate tool sets should be used for water and wastewater operations and be clearly marked or identifiable to distinguish between purposes

- **Storage of equipment**
  - Stores should have procedures in place to prevent any cross contamination between any tools/equipment used for water supply and other uses

- **Training**
  - Any personnel working on restricted operations will need to have undertaken the relevant training and authorised under an approved water hygiene scheme

- **Medical Surveillance**
  - Any personnel working on restricted operations and authorised under an approved water hygiene scheme will have undertaken an initial medical assessment

Further guidance can be found in the Principles of Water Supply Hygiene October 2015

http://www.water.org.uk/publications/reports/principles-water-supply-hygiene

Self-assessment questions

1. Explain briefly what is meant by Commissioning

_____

_____

_____

_____

2. Commissioning of hot and cold water installations includes the following processes, all of which are considered good practice. State which TWO processes are essential requirements of Paragraphs 12 and 13 of Schedule 2.

   Visual inspection

   Flushing and disinfection

   Soundness testing

   Performance testing

   Final checks and handing

   1. _____

   2. _____

3. When looking at those parts of the Guidance document relating to testing, flushing and disinfecting hot and cold water installations, you find references to BS 8558 for more detailed guidance and procedures.

   Underline or highlight which of the following statements is most correct.

   a) BS 8558 is named in Paragraph 12 and its guidance becomes law.

   b) BS 8558 is referred to in the Guidance Document and may be followed for guidance and help only.

   c) BS 8558 is referred to in the Guidance Document and its recommendations may be deemed to satisfy the requirements of Paragraphs 12 and 13.

4. Paragraph 12 sets out criteria for the soundness testing of hot and cold water installations in two different types of material.

   For materials not containing plastics, state:

   a) what the test pressure should be:

   _____

   _____

   b) how long the test period should be: _____

   c) what shows proof of a satisfactory test:

   _____

5. One type of pipe in a system is permitted to be subjected to a slightly less severe test.

   a) name the type of material _____

   b) give one common example of the material _____

   c) give ONE reason why a less severe test may be permitted.

   _____

   _____

   _____

   _____

6. Paragraph 13 gives requirements for the flushing of water systems.

   a) State what pipe or installation should be flushed

   _____

   b) Give any exceptions that are allowed _____

7. Paragraph 13 of Schedule 2 says that after flushing 'systems shall if necessary be disinfected' and goes on to describe the types of installation that are covered by the regulations.

   State TWO situations where disinfection it is NOT considered necessary.

   1. _____

   2. _____

8. Give the FIVE types of installation or situation, mentioned in the Guidance Document, that are expected to be disinfected under Paragraph 13.

   a) _____

   b) _____

   c) _____

   d) _____

   e) _____

9. State the sequence of disinfection for hot and cold water installations

    1. _____

    2. _____

    3. _____

10. Name one common chemical which may be used for the disinfection of water installations and give the solution strength needed for disinfection.

    Common chemical _____

    Solution strength _____

11. Give the four basic stages of the disinfection procedure.

    1. _____

    2. _____

    3. _____

    4. _____

12. There are TWO circumstances where disinfection procedures may need to be repeated. Give both.

    1. _____

    2. _____

13. Give ONE good reason why water should be fed into the system by systematically opening taps and working away from the point of injection (or storage cistern).

    _____

    _____

Check your answers on pages 18, 19 and 20.

Summary of main points

Commissioning means finalising an installation, checking it for faults, putting the system into use, making sure it is to the customers satisfaction and that it operates safely and efficiently.

Commissioning hot and cold water installations includes visual inspection, soundness testing, flushing and disinfection, performance testing, final checks and handing over.

Of these, Soundness testing, flushing and disinfection, are required to be carried out under **Paragraphs 12 and 13 of Schedule 2.**

When soundness testing installations that do not include pipes made of plastics, Paragraph 12(2)(a) sets out **THREE test requirements:**

1. Installations shall be pumped up to a **test pressure of 1.5 times the maximum operating pressure,** plus any expected surge pressure

2. for a test period of one hour

3. during which there should be **no visible leakage** and **no loss of pressure.**

These requirements apply to **all tests** and **all installations.**

For systems containing plastics (elastomeric) pipes or fittings Paragraph 12(2)(b) gives criteria for a less severe a procedure as set out in BS EN 806-4.

Every water system shall, where necessary, be disinfected before it is first used. This means that systems should be disinfected in:

(a) **new installations;**

(b) **major extensions;**

(c) **underground pipework;**

(d) **where contamination may be suspected; or**

(e) **where a system has not been in regular use, and not flushed for 30 days or more.**

The sequence of disinfection should follow the water flow into the premises.

1. water mains

2. service pipes and cisterns

3. distribution system

Flushing and disinfection procedures should follow the recommendations of BS 8558.

For single dwellings and minor alterations and repairs, disinfection will not be necessary, unless contamination is suspected.

Safety

Systems or parts of systems should be taken out of use during any disinfection period, and all outlets marked **DISINFECTION IN PROGRESS – DO NOT USE**.

No other chemicals should be added to the water during disinfection as this may cause generation of toxic fumes.

ALL users of premises should be notified that disinfection is to take place.

Care should be taken when handling disinfection chemicals and safety goggles and clothing worn. Refrain from smoking during disinfection procedures.

Water supply hygiene procedures should follow the recommendations of Principles of Water Supply Hygiene October 2015.

It is essential that all those involved in the production and distribution of drinking water, whether employees of a water undertaker or of a contractor should understand the need to undertake precautions.

Answers to self-assessment questions

1. Explain briefly what is meant by Commissioning

   Commissioning means finalising an installation, checking it for faults, and putting it into use, making sure it is to the customer's satisfaction, and that it operates safely and efficiently.

2. Commissioning of hot and cold water installations includes the following processes, all of which are considered good practice. State which TWO processes are essential requirements of Paragraphs 12 and 13 of Schedule 2.

   Visual inspection

   Flushing and disinfection

   Soundness testing

   Performance testing

   Final checks and handing

   1. ***Flushing and disinfection***
   2. ***Soundness testing***

3. When looking at those parts of the Guidance Document relating to testing, flushing and disinfecting hot and cold water installations, you find references to BS 8558 for more detailed guidance and procedures.

   Underline or highlight which of the following statements is most correct.

   a) *BS 8558 is named in Paragraph 12 and its guidance becomes law.*

   b) *BS 8558 is referred to in the Guidance Document and may be followed for guidance and help only.*

   c) ***BS 8558 is referred to in the Guidance Document and its recommendations may be deemed to satisfy the requirements of Paragraphs 12 and 13.***

4. Paragraph 12 sets out criteria for the soundness testing of hot and cold water installations in two different types of material.

   For materials not containing plastics, state:

   a) *what the test pressure should be:*

       1½ times working pressure plus an allowance for any expected surge

   b) *how long the test period should be:*   ***One hour***

   c) *what shows proof of a satisfactory test:*   ***No visible leakage, and no loss of pressure***

5. One type of pipe is permitted to be subjected to a slightly less severe test.

   a) *Name the type of material*   **plastics or elastomeric**

   b) *Give one common example of the material*   **PVC-U or MDPE**

   c) *Briefly give ONE reason why a less severe test may be permitted.*

   Some plastics pipe materials suffer from stresses that may be retained in the pipe material after the test is over and can lead to pipe failure in the future.

6. Paragraph 13 gives requirements for the flushing of water systems.

   a) *State what pipe or installation should be flushed*   **Every water fitting**

   b) *Give any exceptions that are allowed*   **No exceptions**

7. Paragraph 13 of Schedule 2 says that after flushing, 'systems shall be disinfected' and goes on to describe the types of installation that are covered by the byelaws.

   State TWO situations where disinfection it is NOT considered necessary.

   1. **Private dwellings occupied by a single family**

   2. **Localised repairs or insertion of junctions**

8. State the FIVE types of installation or situation, mentioned in the Guidance Document, that are expected to be disinfected under Paragraph 13.

   a) **New installations**

   b) **Major extensions or alterations**

   c) **Underground pipework**

   d) **Anywhere that contamination may be suspected**

   e) **Where a system has not been in regular use and not regularly flushed.**

9. State the sequence of disinfection for hot and cold water installations

   The sequence should follow the water flow into the premises.

   1. **Water mains**

   2. **Service pipes and cisterns**

   3. **Distribution system**

10. Name one common chemical which may be used for the disinfection of water installations, and give the solution strength needed for disinfection.

  Chemical: **chlorine**

  Solution strength: **50mg/l (50ppm)**

11. Give the four basic stages of the disinfection procedure.

  1. **Thoroughly flush system prior to disinfection**

  2. **Fill system with chlorinated water to concentration of 50mg/l (50ppm) ensuring that the whole system is saturated**

  3. **Leave for one hour then measure for free residual chlorine. If less than 30mg/l the disinfection procedure should be repeated.**

  4. **Thoroughly flush with clean water until residual chlorine is at the same level as the drinking water supplied.**

12. There are TWO circumstances where disinfection procedures may need to be repeated. Give both.

  1. **Where the free residual chlorine is less than 30mg/l after the saturation period,**

  2. **Where the bacterial analysis results are unsatisfactory**

13. Give ONE good reason why water should be fed into the system by systematically opening taps and working away from the point of injection (or storage cistern).

  Filling with water systematically will ensure that the whole system is saturated.

What to do next

Good! You have completed yet another module!

Not too difficult, was it?

So! Go on to **Module 7** Prevention of cross connection to unwholesome water

Water Industry Act 1991:

Water Supply (Water Fittings) Regulations 1999

An Open Learning Course

Module 7

Prevention of cross connection to unwholesome water

Introduction

Water undertakers (suppliers) **have a duty** under the Water Industry Act 1991 **to supply wholesome water**. That is to say, the water they supply to our premises must be clean, free from harmful impurities, and fit for us to drink.

Having been supplied with wholesome water by the undertaker, **it is the duty of installers and users** under the Water Regulations 1999 **not to contaminate the water supplied to us**.

Paragraph 14 looks at one aspect of contamination, namely cross connections, and requires us to make sure that our installations are not connected to any source of water that might cause contamination to any supply pipe or distributing pipe.

What is the requirement?

> *Schedule 2: Paragraph 14:*
>
> *Prevention of cross connection to unwholesome water.*
>
> *14.-(1)* Any water fitting conveying
>
> > (a) rain water, recycled water, or any fluid other than water supplied by the water undertaker; or
> >
> > (b) any fluid that is not wholesome water;
> >
> > shall be clearly identified as to be easily distinguishable from any supply pipe or distributing pipe.
>
> > *(2)* No supply pipe, distributing pipe or pump delivery pipe drawing water from a supply pipe or distributing pipe shall convey, or be connected so that it can convey, any fluid falling within sub-paragraph (1) unless a device for preventing backflow is installed in accordance with Paragraph 15.

What does this requirement mean?

It's quite interesting that two types of water are singled out and mentioned individually in this requirement when it could well have simply referred to 'any fluid that is not wholesome'. That these are given individual attention is a reflection of the government's view that both rainwater and recycled water could make a useful contribution to water conservation in the future.

The Water Regulations Advisory Committee, in its 'recommendations for requirements to replace water byelaws' **suggested that the use of recycled water could make significant contributions to water conservation** and considered that systems making use of such waters are likely to become more widely used in future years. They also made the point that **as the use of these waters is increased, so also will the risk of cross connection and backflow increase.**

Looking at **Paragraph 14.-(1)** we see that it **is concerned with identifying water sources** that might be a contamination risk 'to distinguish them from any supply pipe or distributing pipe'.

So! What water sources are we talking about and how can they be identified?

Well! Apart from recycled water and rainwater there are other waters which are commonly used to provide an alternative or reserve supply of water, to offset the effects of limited pressure, or to provide a supply or storage facility for a specific purpose. These include:

– water supplies and reserve storage facility for firefighting;

– water from private wells, springs etc (not supplied by the undertaker);

– connections between stored water and supply/distributing pipes;

– recycled water from industrial equipment or food producing processes.

We shall look at how these should be connected in a few moments, but first how to identify them.

1. Pipes within buildings should be colour coded to distinguish one from the other this helps the water supply industry, where alternative water reuse systems are becoming more popular – and where alterations can lead to contaminated drinking water if the content of different pipes is not explicit.

BS 1710:2014 will support the legal requirement for those who install and use plumbing systems to prevent cross contamination.

More broadly the standard will be of interest to building designers, operators, users and service installers, and a wide variety of associated industries and activities including waste, water, liquid fuels, gases and refrigeration".

Currently, it is generally only the larger industrial and commercial buildings that have their pipes colour coded, but where a house or other small building uses water other than potable water as supplied by the undertaker, there is a case for colour coding in these buildings also. The colour codes for drinking water pipes and distributing pipes are shown below.

Since 1984, water services within commercial and industrial buildings have had some form of identification applied to them using BS 1710 – Specification for identification of pipelines and services. Since the standard was introduced, buildings services have changed as have the legal requirements within water legislation. In December 2014, the standard was revised bringing it up to date and enabling users to more easily meet the requirements of Water Supply (Water Fittings) Regulations and Scottish Water Byelaws.

The new standard uses a simpler system of colours and information labels to identify the basic nature and quality of fluids being transported and the applications being served. To meet this requirement there are three key elements which need to be followed:

- Coloured identification bands
- Additional information highlighting the designated usage, pressures and qualities
- Identification of flow (Directional arrow)

It's not only water that can cause contamination and Requirement 14 recognises this and talks about fluids, not waters.

The requirement of Paragraph 14(1) is that pipes conveying unwholesome water must be easily distinguished from supply pipes and distributing pipes.

For information regarding the colours and the widths of the basic, safety, code colour bands, appropriate colour number and the dimension requirements refer to BS 1710 and BS 4800.

Example of pipe identification

Colour identification banding ⌐

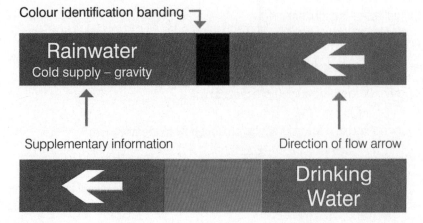

Supplementary information   Direction of flow arrow

Pipe colour guide

Examples of the new water services colour requirements. Pipes that predominately contain water shall be coloured green, with an additional colour applied to identify the source water. Auxiliary Blue is applied for systems connected to the public water supply and Flint Grey for water from any other source such as well, boreholes or water re-use systems.

Typical application	Basic colour identification band				Basic colour identification band
Source of water Potable designation for water meeting drinking water standards					Water provided from the public supply (i.e. water undertaker)
					Water derived from a source other than the public supply (i.e. private borehole, well etc.)
End use water quality An additional black band to be applied where the end use fluid is not intended to meet standards for drinking water					Public water supply system
					Any other water source
Safety systems Fire systems connected to a drinking water mains and containing no additives, following an assessment, may be considered for potable designation					Public water supply system
					Any other source
Non-potable designation to be applied to fire systems, which are fed from a dedicated fire storage cistern, containing additives or where there is doubt regarding the water quality					Public water supply system
					Any other source

Colour coding key

Green ▮ Black ▮ Flint Grey ▮ Red ▮ Auxiliary Blue ▮

2. Secondly, identification can be made at draw-off points and **taps should be labelled to identify those that supply water for drinking purposes and those that are not suitable for drinking.**
Remember! Paragraph 14(1) requires fittings containing unwholesome water to be identified.

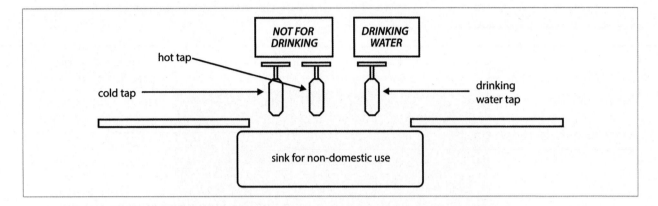

3. Thirdly, **accurate layout drawings should be made and passed over to the customer showing where pipes are situated both within buildings and below ground** so as to help distinguish them from one another.

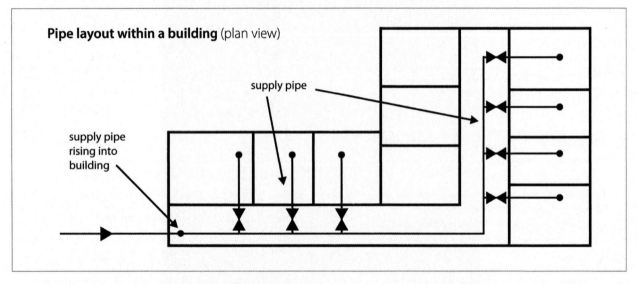

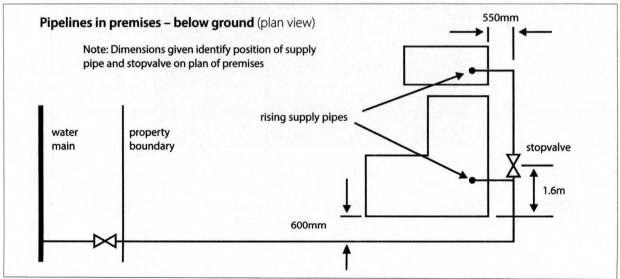

How should we prevent or avoid making cross connections?

The first step is to **recognise the potential hazard** and we have looked at that already. Secondly, we should make sure we **connect the pipe runs so that the hazard is avoided**. We have listed previously a number of liquids that should not be connected to supply pipes and distributing pipes. These are shown in a number of diagrams illustrating both good and bad practice.

It is also important to remember that supply pipes and distributing pipes should not be connected together, (even though they may both be supplying drinking water) unless in certain cases (e.g. shower mixers) appropriate backflow devices are fitted.

We shall look at backflow in the next module.

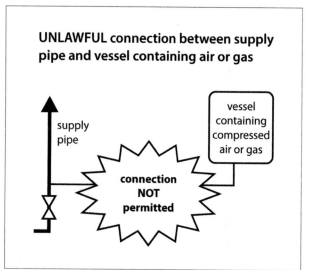

UNLAWFUL connection between supply pipe and vessel containing air or gas

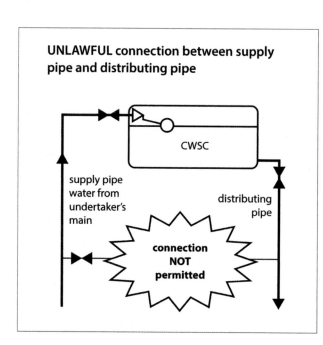

UNLAWFUL connection between supply pipe and distributing pipe

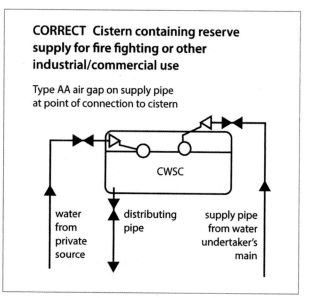

CORRECT Cistern containing reserve supply for fire fighting or other industrial/commercial use

Type AA air gap on supply pipe at point of connection to cistern

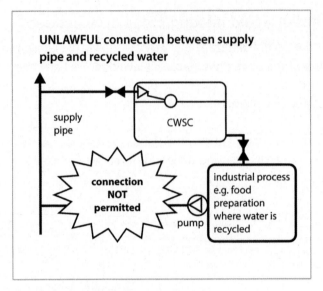

UNLAWFUL connection between supply pipe and recycled water

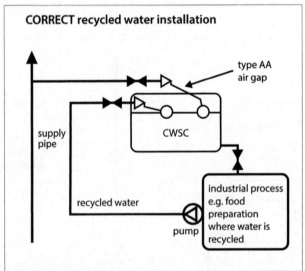

CORRECT recycled water installation

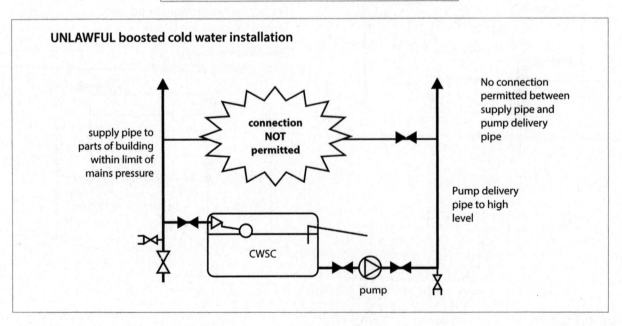

UNLAWFUL boosted cold water installation

Self-assessment questions

1. Water Authorities have a duty under the Water Industry Act 1991 to supply wholesome water to its customers. What similar duty is placed on the installer or contractor under the Water Supply (Water Fittings) Regulations 1999?

_____

2. It is suggested that in the future rainwater and recycled water will make a contribution to water conservation. Suggest what effect the increased use of these waters will have in terms of contamination of wholesome water.

_____

3. Requirement 14.-(1) requires us to identify 'alternative' water sources and distinguish them from any supply pipe or distributing pipe. State why this should be done.

_____

_____

4. One way of identifying pipes within buildings, and distinguishing them one from the other, is by 'colour coding' them in accordance with the recommendations of BS 1710. Indicate the recommended colour codings for the pipes illustrated in the following diagram.

Colour coding for supply pipe and distributing pipes

Water provided from the public supply (i.e. water undertaker)			
Basic identification colour	Safety and code colours		Basic identification colour
_____	_____		_____

Public water supply system (end use water quality)				
Basic identification colour	Safety and code colours	Safety and code colours	Safety and code colours	Basic identification colour
_____	_____	_____	_____	_____

Any other source (end use water quality)				
Basic identification colour	Safety and code colours	Safety and code colours	Safety and code colours	Basic identification colour
_____	_____	_____	_____	_____

Public water supply system (safety systems)				
Basic identification colour	Safety and code colours	Safety and code colours	Safety and code colours	Basic identification colour
_____	_____	_____	_____	_____

5. There are TWO other ways in which we can identify pipes one from the other, particularly in larger and more complex buildings. State what they are.

1. _____ 2. _____

6. State the identification colour given to water pipes below ground to distinguish them from other services such as gas electricity etc.

_____

7. The FOUR following diagrams show examples of correct and incorrect installations. Indicate in the boxes provided which diagram show correct installation and which is incorrect. Also, state why the incorrect installations would not be permitted.

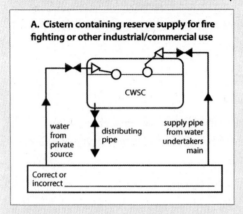

A. Cistern containing reserve supply for fire fighting or other industrial/commercial use

CWSC

water from private source   distributing pipe   supply pipe from water undertakers main

Correct or incorrect _____

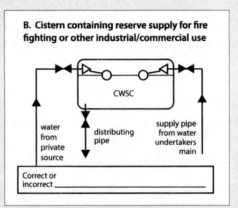

B. Cistern containing reserve supply for fire fighting or other industrial/commercial use

CWSC

water from private source   distributing pipe   supply pipe from water undertakers main

Correct or incorrect _____

Diagram _____ would not be permitted because _____

_____

_____

_____

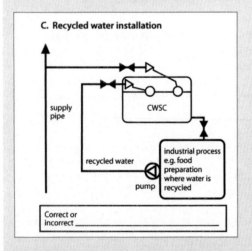

C. Recycled water installation

CWSC

supply pipe

recycled water

pump   industrial process e.g. food preparation where water is recycled

Correct or incorrect _____

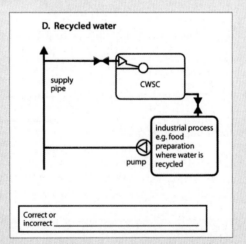

D. Recycled water

supply pipe

CWSC

industrial process e.g. food preparation where water is recycled

pump

Correct or incorrect _____

Diagram _____ would not be permitted because _____

_____

_____

_____

Check your answers on page 12 and 13.

Summary of main points

Water undertakers (suppliers) **have a duty to supply wholesome water, and it is the duty of installers and users not to contaminate the water supplied to us.**

Requirement 14.-(1) is concerned with identifying water sources that might be a contamination risk **'to distinguish them from any supply pipe or distributing pipe'.**

1. **Pipes within buildings should be colour coded** in accordance with BS 1710 which gives guidance for a wide variety of piped services within buildings.

   Remember! **It's not only water that can cause contamination!** Requirement 14 recognises this and talks about protection from fluids, not just waters.

   The requirement of Paragraph 14(1) is that **pipes conveying unwholesome water must be easily distinguished from supply pipes and distributing pipes.**

2. **Taps should be labelled to show which tap is suitable for drinking** purposes and those that are not suitable for drinking.

3. **Drawings should be made and passed over to the customer showing where pipes are situated both within buildings and below ground.**

The first step in **crossflow protection** is to **recognise the potential hazard,** then, **connect pipe runs so that the hazard is avoided.**

Supply pipes should not be **directly** connected to:
- pipes containing water from any other source;
- distributing pipes;
- recirculating pipes;
- pumped delivery pipes;
- vessels containing air or gas;
- or any other source of contamination

unless appropriate backflow prevention devices, which may be an air gap or interposed cistern, are in place.

Please note that a full explanation of backflow devices is covered in Module 8.

Answers to self-assessment questions

1. Water Authorities have a duty under the Water Industry Act 1991 to supply wholesome water to its customers. What similar duty is placed on the installer or contractor under the Water Supply (Water Fittings) Regulations 1999?

   The installer is required not to contaminate water supplied by the undertaker.

2. It is suggested that in the future rainwater and recycled water will make a contribution to water conservation. Suggest what effect the increased use of these waters will have in terms of contamination of wholesome water.

   It will increase the risk of contamination through cross connection.

3. Requirement 14.-(1) requires us to identify 'alternative' water sources and distinguish them from any supply pipe or distributing pipe. State why this should be done.

   Because any water NOT supplied by the water undertaker, and any water that has been used or treated in any way, is considered a contamination risk.

4. One way of identifying pipes within buildings, and distinguishing them one from the other, is by 'colour coding' them in accordance with the recommendations of BS 1710. Indicate the recommended colour codings for the pipes illustrated in the following diagram.

Colour coding for supply pipe and distributing pipes

Water provided from the public supply (i.e. water undertaker)		
Basic identification colour **Green**	Safety and code colours **Auxiliary blue**	Basic identification colour **Green**

Public water supply system (end use water quality)				
Basic identification colour **Green**	Safety and code colours **Auxiliary blue**	Safety and code colours **Black**	Safety and code colours **Auxiliary blue**	Basic identification colour **Green**

Any other source (end use water quality)				
Basic identification colour **Green**	Safety and code colours **Flint grey**	Safety and code colours **Black**	Safety and code colours **Flint grey**	Basic identification colour **Green**

Public water supply system (safety systems)				
Basic identification colour **Green**	Safety and code colours **Auxiliary blue**	Safety and code colours **Red**	Safety and code colours **Auxiliary blue**	Basic identification colour **Green**

5. There are TWO other ways in which we can identify pipes one from the other, particularly in larger and more complex buildings. State what they are.

   1. Label taps   *2. Make detailed drawings*

6. Give the identification colour given to water pipes below ground to distinguish them from other services such as gas electricity etc.

   Blue

7. The FOUR diagrams shown below illustrate examples of correct and incorrect installations. Indicate in the boxes provided which diagram show correct installation and which is incorrect. Also, state why the incorrect installations would not be permitted.

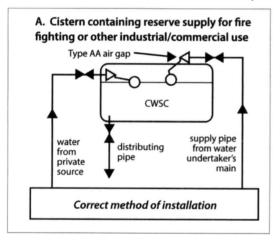

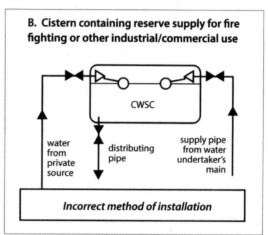

*Diagram **B** would not be permitted because* **there is a possibility that backflow could occur from the cistern to contaminate the supply pipe.**

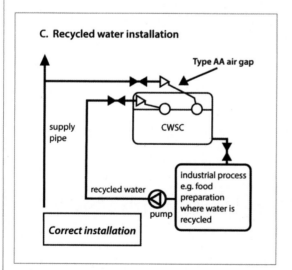

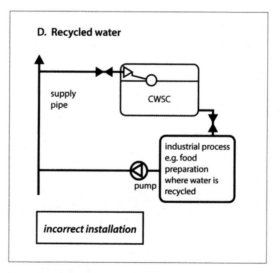

*Diagram **D** would not be permitted because* **the water in the supply pipe could become contaminated by the used water pumped from the industrial process.**

What to do next

Well done! This is another module completed.

Now you should look at backflow which is perhaps the most important subject area dealing with the prevention of contamination of water.

Backflow is also one of the largest modules you have to cope with in this package.

Keep up the good work, and go on to Module 8

Water Industry Act 1991:

Water Supply (Water Fittings) Regulations 1999

An Open Learning Course

Module 8

Backflow prevention

Introduction to backflow prevention

Backflow can be described as *'...the flow of water in a direction contrary to the intended normal direction of flow...'*, and **backflow prevention** is perhaps the most important element of the Water Regulations and its requirements.

It is important that water supplied by the water undertaker to a premises or a process, is not allowed to return to the undertaker's main because of the possibility that it may become contaminated after being supplied.

Water can become contaminated in a number of ways, simply by 'normal' use within a premises, or by misuse or incorrect installation. The degree of contamination will vary according to the particular process and the severity of contaminants with which the water may come into contact.

Water is supplied by the water undertaker in wholesome (drinkable) condition. It is supplied for domestic purposes and it is also supplied for industrial, commercial, agricultural and other processes.

No one would like to think that water used to mix with chemicals in an industrial or agricultural process could possibly 'flow back' into the water main to be supplied to someone else for drinking or food preparation.

In the domestic scene, how would we feel if our bath water was allowed to 'backflow' into the supply pipe to be drawn off later to make our tea?

Obviously these scenarios depict situations that could affect, and perhaps endanger our health.

For these important health reasons, Water Regulations follow previous Water Byelaws in prescribing methods of protection against backflow from our water supply systems.

Incidentally, the term domestic purposes was used above. If you are not aware of its meaning, why not look it up in the accompanying 'Glossary of terms'?

What is the requirement?

> ### Schedule 2: Paragraphs 15 Backflow prevention
>
> **15-(1)** *Subject to the following provisions of this paragraph, every water system shall contain an adequate device or devices for preventing backflow of fluid from any appliance, fitting or process from occurring.*
>
>   **(2)** *Paragraph (1) does not apply to:*
>
>   *(a) a water heater where the expanded water is permitted to flow back into a supply pipe; or*
>
>   *(b) a vented water storage vessel supplied from a storage cistern, where the temperature of the water in the supply pipe or the cistern does not does not exceed 25°C.*
>
>   **(3)** *The device used to prevent backflow shall be appropriate to the highest applicable fluid category to which the fitting is subject downstream before the next such device.*
>
>   **(4)** *Backflow prevention shall be provided on any supply pipe or distributing pipe:*
>
>   *(a) where it is necessary to prevent backflow between separately occupied premises, or*
>
>   *(b) where the water undertaker has given written notice for the purpose of this Schedule above that such prevention is needed for the whole or part of any premises.*
>
>   **(5)** *A backflow prevention device is adequate for the purpose of Paragraph (1) if it is in accordance with a specification approved by the regulator for the purpose of this Schedule.*

You may have noticed that Paragraph 15(5) refers to '**...a specification approved by the regulator...**'. Let's deal with this first.

Because of the importance of backflow prevention, the Secretary of State for the Department of the Environment, Food and Rural Affairs (DEFRA) has decided that the requirements of Paragraph 15 need to be enhanced by a more detailed set of requirements that will provide better understanding of backflow prevention requirements and should, at the same time, achieve higher standards in the implementation of them.

This additional document is the **Regulators' Specifications on the Prevention of Backflow which is enforceable from 1 May 2000**.

Prior to 1 May 2000 the backflow requirements of previous Water Byelaws were deemed to satisfy the requirements of the Regulator's Specification.

The Regulator's Specification describes a range of backflow prevention devices, and provides essential information on the application and installation of them. The contents of the Regulator's Specification are reflected in this training module.

What is backflow?

Backflow is defined in Paragraph 1 of Schedule 2 as *'flow upstream, that is in a direction contrary to the intended normal direction of flow, within or from a water fitting;'*

More simply, backflow can be defined as *'the flow of water in the opposite direction to that which was intended'*.

Backflow can be caused by **backpressure** or **backsiphonage**.

Backpressure can be described as *'the reversal of flow in a pipe caused by an increase in pressure in the system'*.

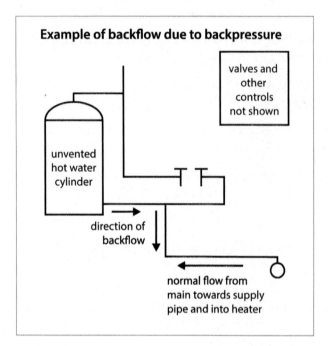

Example of backflow due to backpressure

valves and other controls not shown

unvented hot water cylinder

direction of backflow

normal flow from main towards supply pipe and into heater

An example of backpressure can be seen in the expansion of water from an unvented hot water heater which is permitted to pass back into the supply pipe due to the expansion of water in the hot store vessel when heated.

The example shown **may** be permitted in some circumstances. (See Module 10) but the expanded water must not be permitted to reach a point where the water could be drawn off from the supply pipe.

Backsiphonage is *'backflow caused by siphonage of water from a cistern or appliance back into the pipe which feeds it'*.

Example: hosepipe in use when mains are turned off or severe break in the main occurs causing water to be siphoned back along a supply pipe towards the main.

Note: Further examples of backflow device are shown in the Glossary of terms which accompanies this package.

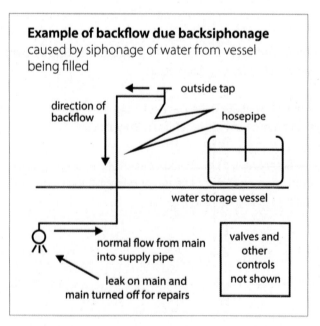

Example of backflow due backsiphonage caused by siphonage of water from vessel being filled

outside tap

direction of backflow

hosepipe

water storage vessel

normal flow from main into supply pipe

valves and other controls not shown

leak on main and main turned off for repairs

What are backflow prevention devices?

A backflow prevention device is defined as *'a device which is intended to prevent contamination of drinking water by backflow'*. It is quite simply either a mechanical or non-mechanical fitting or arrangement within a pipework system that is there to stop backflow from occurring.

There are quite a number of backflow prevention devices in various shapes and forms. Typical examples are the check valve, the double check valve and the various air gap arrangements. You will probably have become familiar with these from past water byelaw requirements, although some devices have been given new names under the requirements of the Regulations and its accompanying regulator specification.

Water Regulations now list **10 non-mechanical devices** and **14 mechanical devices**. The list, which also indicates the risk category, or severity of the pollution risk, for which each device is suited, is shown in Tables S6.1 and S6.2 of the regulator's specification, and reproduced here on pages 8 and 9.

We shall be looking at these devices in more detail later, but first we need to look at the **fluid risk categories**.

What are the fluid risk categories?

Over the past 10 to 12 years we have become familiar with three categories of backflow risk required by Water Byelaws which implemented the recommendations of the Backsiphonage report of 1974.

Schedule 1 of the Water Regulations now recognises and implements a more comprehensive range of **five fluid risk categories** based on that developed by the Union of Water Supply Associations of Europe (EUREAU 12) and that is also used in North America and Australia.

Water Byelaws – risk categories	Water Regulations – fluid risk categories
	5 Serious health hazard
1 Serious health hazard	**4** Significant health hazard
2 Significant health hazard	**3** Slight health hazard
3 Aesthetic quality is impaired	**2** Aesthetic quality is impaired
Comparison of classes of risk with fluid risk categories	**1** No health hazard or impairment in its quality

The five fluid risk categories represent various qualities of water depending on how 'drinkable' they are, or how much danger to health they might be, which in turn depends on what impurities they contain.

The five fluid categories, taken from Schedule 1 are reproduced below:

Fluid category 1

Wholesome water supplied by a water undertaker and complying with the requirements of regulations made under Section 67 of the Water Industry Act 1991(a).

Fluid category 2

Water in fluid category 1 whose aesthetic quality is impaired owing to:

(a) a change in its temperature, or

(b) the presence of substances or organisms causing a change in its taste, odour or appearance, including water in a hot water distributing system

Fluid category 3

Fluid which represents a slight health hazard because of the concentration of substances of low toxicity, including any fluid which contains:

(a) ethylene glycol, copper sulphate solution or similar chemical additives, or

(b) sodium hypochlorite (chloros and common disinfectants)

Fluid category 4

Fluid which represents a significant health hazard because of the concentration of toxic substances, including any fluid which contains:

(a) chemical, carcinogenic substances or pesticides (including insecticides and herbicides), or

(b) environmental organisms of potential health significance

Fluid category 5

Fluid representing a serious health hazard because of the concentration of pathogenic organisms, radioactive or very toxic substances. including any fluid which contains:

(a) faecal matter or other human waste

(b) butchery or other animal waste, or

(c) pathogens from any other source

So now you have some idea of the relationship between backflow risk and the five fluid categories.

Do we need to know about backflow prevention devices?

Yes you do! You need to know exactly which backflow device is needed for a particular installation, and to help make that decision, you need to know which fluid risk category the device is to be used to give protection from.

These two points are all sorted out for us in the **Regulators' Specification on the Prevention of Backflow** where a number of tables are provided:

(a) Tables **S6.1 and S6.2** provide a schedule of mechanical and non-mechanical backflow devices, and tells us the fluid category for which each device is suited.

(b) Tables **6.1(a) to (e)** help us decide which fluid category a fluid is in.

The above tables are reproduced in this module on pages 8 to 12.

For example, let us assume you have to fit a base exchange water softener in a commercial or industrial building.

By referring to the tables on pages 10, 11 and 12. you will see commercial water softeners listed as a **fluid category 3 (low toxicity) risk**. Check it out for yourself!

Now move to the table on pages 8 and 9 where you will find a number of backflow prevention devices for category 3 risks. In this case one that will guard against both backpressure and backsiphonage, and which is suited to an in-line application is a double check valve listed in the regulations as:

* **Type EC Verifiable double check valve, or**

* **Type ED Non-verifiable double check valve**

When looking this up, be careful not to mistake the commercial water softener with the domestic water softener in Table 3.2 because a softener used in a house is considered a lesser risk than one for commercial/industrial purposes and will not need the same backflow protection device.

Why not look up the device needed for a domestic water softener and see just what the difference is?

Is that all there is to it?

No! Not at all!

Once you have decided on the fluid category and sorted out which device is needed you will then need to look at any installation requirements that apply.

We shall deal with that as we go through! But first let's look at the tables referred to earlier.

Table S6.1 Schedule of non-mechanical backflow prevention arrangements and the maximum permissible fluid category for which they are acceptable

Type	Description of backflow prevention arrangements and devices	Fluid category for which suited	
		Back-pressure	Back-siphonage
AA	Air gap with unrestricted discharge above spillover level	5	5
AB	Air gap with weir overflow	5	5
AC	Air gap with vented submerged inlet	3	3
AD	Air gap with injector	5	5
AF	Air gap with circular overflow	4	4
AG	Air gap with minimum size circular overflow determined by measure or vacuum test	3	3
AUK1	Air gap with interposed cistern (for example, a WC suite)	3	5
AUK2	Air gaps for taps and combination fittings (tap gaps) discharging over domestic sanitary appliances, such as a wash basin, bidet, bath or shower tray shall not be less than the following: **size of tap or combination fitting** **vertical distance of bottom of tap outlet above spillover level of receiving applicant** not exceed G½ 20mm exceeding G½ but not exceeding G¾ 25mm exceeding G¾ 70mm	X	3
AUK3	Air gaps for taps or combination fittings (tap gaps) discharging over any higher risk domestic sanitary appliances where a fluid category 4 or 5 is present, such as: any domestic or non-domestic sink or other appliance; or any appliance in premises where a higher level of protection is required, such as some appliances in hospitals or other health care premises, shall be not less than 20mm or twice the diameter of the inlet pipe to the fitting, whichever is the greater	X	5
DC	Pipe interrupter with permanent atmospheric vent	X	5

Notes:

1. X indicates that the backflow prevention arrangement or device is not applicable or not acceptable for protection against backpressure for any fluid category within water installations in the UK.

2. Arrangements incorporating Type DC devices shall have no control valves on the outlet side of the device; they shall be fitted not less than 300mm above the spillover level of a WC pan, or 150mm above the sparge pipe outlet of a urinal, and discharge vertically downwards.

3. Overflows and warning pipes shall discharge through, or terminate with, an air gap, the dimension of which should satisfy a Type AA air gap.

Table S6.2 Schedule of mechanical backflow prevention arrangements and the maximum permissible fluid category for which they are acceptable

Type	Description of backflow prevention arrangements and devices	Fluid category for which suited	
		Back-pressure	Back-siphonage
BA	Verifiable backflow preventer with reduced pressure zone	4	4
CA	Non-verifiable disconnector with difference between pressure zones not greater than 10%	3	3
DA	Anti-vacuum valve (or vacuum breaker)	X	3
DB	Pipe interrupter with atmospheric vent and moving element	X	4
DUK1	Anti-vacuum valve combined with a single check valve	2	3
EA	Verifiable single check valve	2	2
EB	Non-verifiable single check valve	2	2
EC	Verifiable double check valve	3	3
ED	Non-verifiable double check valve	3	3
HA	Hose Union backflow preventer. Only permitted for use on existing hose union taps in house installations.	2	3
HC	Diverter with automatic return (normally integral with some domestic appliance applications only)	X	3
HUK1	Hose union tap which incorporates a double check valve. Only permitted for replacement of existing hose union taps in house installations.	3	3
LA	Pressurised air inlet valve	X	2
LB	Pressurised air inlet valve combined with a check valve downstream	2	2

Notes:

1. X indicates that the backflow prevention device is not acceptable for protection against backpressure for any fluid category within water installations in the UK.

2. Arrangements incorporating a Type DB device shall have no control valves on the outlet side of the device. The device shall be fitted not less than 300mm above the spillover level of an appliance and discharge vertically downwards.

3. Type DA and DUK1 shall have no control valves on the outlet side of the device and be fitted on a 300mm minimum Type A upstand.

4. Relief outlet ports from Types BA and CA backflow prevention devices shall terminate with an air gap, the dimension of which should satisfy a Type AA air gap.

Determination of fluid categories with examples

Table 6.1a: Determination of fluid category 1

Fluid category 1: *Wholesome water supplied by a water undertaker and complying with the requirements of regulations made under Section 67 of the Water Industry Act 1991(a).*

Example *Water supplied directly from a water undertaker's main*

Table 6.1b: Determination of fluid category 2

Fluid category 2: *Water in fluid category 1 whose aesthetic quality is impaired owing to:*

(a) *a change in its temperature, or*

(b) *the presence of substances or organisms causing a change in taste, odour or appearance, including water in a hot water distribution system*

Examples *Mixing of hot and cold water supplies*

Domestic softening plant (common salt regeneration)

Drink vending machine in which no ingredients or carbon dioxide are injected into the supply or distributing inlet pipe

Fire sprinkler systems (without anti-freeze)

Ice making machines

Water cooled air conditioning units (without additives)

Table 6.1c: Determination of fluid category 3

Fluid category 3: *Fluid which represents a slight health hazard because of the concentration of substances of low toxicity, including any fluid which contains:*

(a) *ethylene glycol, copper sulphate solution or similar chemical additives; or*

(b) *sodium hypochlorite (chloros and common disinfectants)*

Examples *Water in primary circuits and heating systems (with or without additives) in a house*

Domestic wash-basins, baths and showers

Domestic clothes and dish washing machines

Home dialysing machines

Drink vending machines in which ingredients or carbon dioxide are injected

Commercial softening plant (common salt regeneration)

Domestic hand held hoses with flow controlled spray or shut-off control

Hand held fertiliser sprays for use in domestic gardens

Domestic or commercial irrigation systems, without insecticide or fertiliser additives, and with fixed sprinkler heads not less than 150 mm above ground level

Determination of fluid categories with examples

Table 6.1d: Determination of fluid category 4

Fluid category 4: *Fluid which represents a significant health hazard because of the concentration of toxic substances, including any fluid which contains:*

    (a) *chemical, carcinogenic substances or pesticides (including insecticides and herbicides), or*

    (b) *environmental organisms of potential health significance.*

Examples

General
Primary circuits and central heating circuits in other than a house
Fire sprinkler systems using antifreeze solutions

House gardens
Mini-irrigation systems without fertiliser or insecticide application;
such as pop-up sprinklers or permeable hoses

Food processing
Food preparation
Dairies
Bottle washing apparatus

Catering
Commercial dish washing machines
Bottle washing apparatus
Refrigerating equipment

Industrial and commercial installations
Dying equipment
Industrial disinfection equipment
Printing and photographic equipment
Car washing and degreasing plants
Commercial clothes washing plants
Brewery and distillation plant
Water treatment or softeners using other than salt
Pressurised fire-fighting systems

Table 6.1e: Determination of fluid category 5

Fluid category 5: *Fluid which represents a serious health hazard because of the concentration of pathogenic organisms, radio active or very toxic substances, including any fluid which contains:*

(a) *faecal material or other human waste;*

(b) *butchery or other animal waste; or*

(c) *pathogens from any other source.*

Examples

General

Industrial cisterns

Non-domestic hose union taps

Sinks, Urinals, WC pans and bidets

Permeable pipes in other than domestic gardens, laid at or below ground level, with or without chemical additives

Grey water recycling systems

Medical

Any medical or dental equipment with submerged inlets

Laboratories

Bedpan washers

Mortuary and embalming equipment

Hospital dialysis machines

Commercial clothes washing plant in health care premises

Non-domestic sinks, baths, wash-basins and other appliances

Food processing

Butchery and meat trades

Slaughterhouse equipment

Vegetable washing

Catering

Dish washing machines in health care premises

Vegetable washing

Industrial and commercial installations

Industrial and commercial plant etc

Mobile plant, tankers and gully emptiers

Laboratories

Sewage treatment and sewage cleansing

Drain cleaning plant

Water storage for agricultural purposes

Water storage for fire-fighting purposes

Commercial agricultural

Commercial irrigation outlets below or at ground level and/or permeable pipes, with or without chemical additives

Insecticide or fertiliser applications

Commercial hydroponic systems

Note: *The list of examples shown above for each fluid category is not exhaustive*

What are backflow devices and where should they be positioned?

Most backflow devices are 'point of use' devices, which means they are used to protect against each individual risk at, or near the point of supply, or where the actual risk is likely to occur. Backflow protection may be achieved by using one or more of the devices previously listed, and the type of device used will depend on the severity of the risk.

For example, if, on a farm, water is used to mix chemicals for the spraying of crops with fertilisers, it is obviously important that no water from the contaminated storage vessel can get back into the supply pipe that feeds it. So! To prevent this, a backflow prevention device must be used, and **positioned at or near to the point where the water is supplied to the vessel.** Hence the term 'point of use' device.

In this case, for a commercial or agricultural application, a suitable device would be a Type AA air gap or Type AB air gap with weir overflow, or some other no less effective device

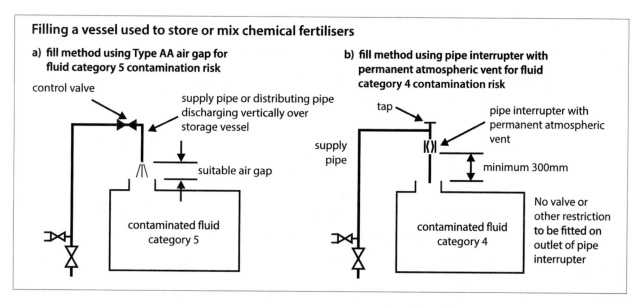

Filling a vessel used to store or mix chemical fertilisers

a) **fill method using Type AA air gap for fluid category 5 contamination risk**

control valve
supply pipe or distributing pipe discharging vertically over storage vessel
suitable air gap
contaminated fluid category 5

b) **fill method using pipe interrupter with permanent atmospheric vent for fluid category 4 contamination risk**

tap
supply pipe
pipe interrupter with permanent atmospheric vent
minimum 300mm
contaminated fluid category 4
No valve or other restriction to be fitted on outlet of pipe interrupter

Another example! If a supply pipe carrying water from the main into a building for general and domestic use is to have a branch supply for a sprinkler system, water from the sprinkler system (fluid category 2) must not be permitted to flow back into the supply pipe (fluid category 1). By reference to the tables on page 10 we can see that a verifiable or non-verifiable single check valve will be suitable, which must be fitted at the branch to the sprinkler pipeline.

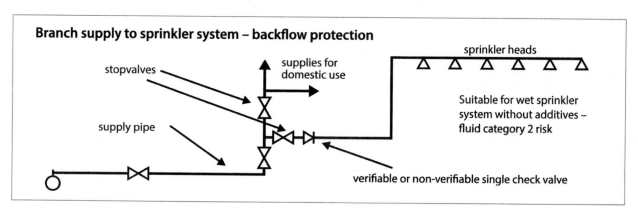

Branch supply to sprinkler system – backflow protection

sprinkler heads
stopvalves
supplies for domestic use
supply pipe
Suitable for wet sprinkler system without additives – fluid category 2 risk
verifiable or non-verifiable single check valve

What about the installation of backflow prevention devices?
Are there any special rules that apply?

In addition to what we have already discussed, the Regulators' Specification on the Prevention of Backflow gives the following general advice on the installation of backflow devices:

G15.2. **Backflow problems can be avoided by good system design and the provision of suitable backflow devices.**

G15.5. **Mechanical backflow devices should be avoided where practicable by the use of air gaps (or tap gaps) at the point of use.**

G15.6. **Permanently vented distributing pipes will provide good 'secondary' protection (whole site or zone protection) in many cistern fed installations.**

G15.7. **Backflow prevention devices:**

    (a) **should be readily accessible for inspection, operational maintenance and renewal.**

    (b) **for risk categories 2 and 3, devices should NOT be located outside the building,** (except types HA and HUK1 which are meant for existing outside taps).

    (c) **should not be buried in the ground.**

    (d) of the **vented or verifiable type should not be installed in chambers below ground level or where liable to flooding.**

    (e) used for **category 4 devices, should have line strainers fitted upstream** (before the backflow device) **and a servicing valve upstream of the strainer.**

    (f) of the **reduced pressure zone valve type, the relief outlet ports should terminate with a Type AA air gap located at least 300mm above ground or floor level.**

Many of these points will be looked at again when dealing with individual devices or their application to installations.

Next we look at the individual devices, but first some questions on what we have discussed so far.

Please go to the next page!

Self-assessment questions (1)

1. Give a simple description of backflow

   Backflow can be described as _____

   _____

2. The following two diagrams illustrate backpressure and backsiphonage. Complete the titles to show which type of backflow is represented.

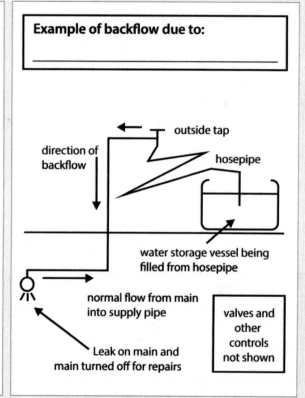

Example of backflow due to:

_____

valves and other controls not shown

unvented hot water cylinder

direction of backflow

normal flow from main towards supply pipe and into heater

Example of backflow due to:

_____

outside tap

direction of backflow

hosepipe

water storage vessel being filled from hosepipe

normal flow from main into supply pipe

valves and other controls not shown

Leak on main and main turned off for repairs

3. Describe backpressure and backsiphonage

   Back-pressure *can be described as* _____

   _____

   Backsiphonage *is* _____

   _____

4. Schedule 1 of the Water Supply (Water Fittings) Regulations 1999 sets out five fluid categories which are shown in the following chart. Complete the chart to show the severity of risk in each case.

Water Regulations – fluid risk categories

5 _____ _health hazard_

4 _____ _health hazard_

3 _____ _health hazard_

2 _____

1 _____
   _____

5. Most backflow devices are 'point of use' devices. Explain briefly what is meant by a 'point of use' device?

_____

_____

6. Complete the following statements giving general advice on the installation of backflow prevention devices.

G15.7. Backflow prevention devices:

a) *should be* _____ _____ *for inspection,*
   operational maintenance and renewal

b) *for risk categories 2 and 3, except Types HA and HUK1, devices should* _____
   be positioned outside the building

c) *should not be buried in the* _____

d) *of the vented or verifiable type should NOT be installed in chambers*
   _____ _____ *or where liable to* _____

e) *used for category 4 devices, should have* _____ _____ *fitted upstream*
   and a _____ _____ *upstream of the strainer.*

f) *of the reduced pressure zone valve type, should terminate with a type* _____ *air gap*
   located at least _____ *above ground or floor level.*

Check your answers on pages 67 and 68.

Non-mechanically operated backflow prevention devices

Air gaps. The Regulator's Specification refers to seven different air gaps. Each is allotted to one of three fluid categories of risk, 3, 4 or 5 depending on how well it operates to protect the supply. It is a well known fact that air gaps, being non-mechanical devices will continue to work well with little or no maintenance providing they are set up correctly in the first place.

Type AA Air with unrestricted discharge
(formerly known as Type A air gap)

Means a non-mechanical backflow prevention arrangement of water fittings where the water is discharged through an air gap into a receptacle which has at all times an unrestricted spillover level to the atmosphere.

Application of device: Backpressure – category 5

Backsiphonage – category 5

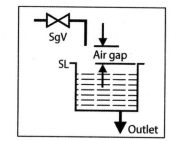

Examples of use: Cold water storage vessel containing fluid that is a serious health risk such as sinks in domestic or industrial/commercial premises, or cisterns for industrial use or for agricultural purposes or water storage for firefighting

---

Type AA Air gap to protect supply pipe delivering water to a cistern or vessel used to store or mix contaminated liquid

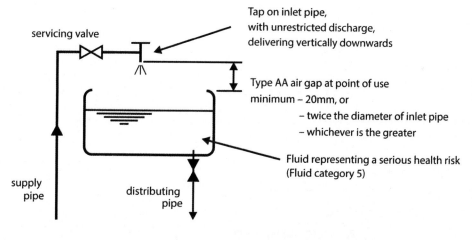

Tap on inlet pipe, with unrestricted discharge, delivering vertically downwards

Type AA air gap at point of use
minimum – 20mm, or
– twice the diameter of inlet pipe
– whichever is the greater

Fluid representing a serious health risk
(Fluid category 5)

servicing valve

supply pipe

distributing pipe

Tap at kitchen sink

Minimum air gap is not normally a problem on kitchen sinks where space is allowed to fill vessels e.g. buckets

Type AUK3 air gap to at point of use

kitchen sink is considered to be a fluid category 5 backflow risk

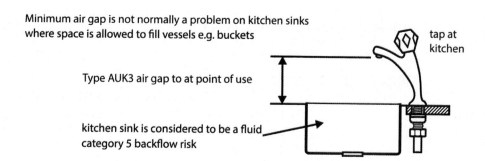

tap at kitchen

Type AB Air gap with weir overflow

Means a non-mechanical backflow prevention arrangement of water fittings complying with Type AA except that the air gap is the vertical distance from the lowest point of discharge orifice which discharges into the receptacle, to the critical level of the rectangular weir overflow.

Application of device: Backpressure – category 5
Backsiphonage – category 5

Examples of use: cold water storage vessel containing fluid that is a serious health risk such as agricultural cattle drinking trough or feed and expansion cistern to commercial or industrial premises.

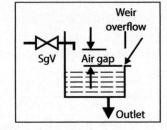

Type AB Air gap on water supply to cattle trough

Float operated valve encased in lockable chamber

Servicing valve

Stopvalve

Air gap from outlet of float valve to spillover level:
– 20mm minimum, or
– twice the diameter of inlet pipe,
– whichever is the greater

Type AB air gap with weir overflow

cattle trough

Pipe insulated and covered to protect from moisture and damage

Type AB Air gap to feed and expansion cistern in commercial or industrial premises
Also suitable for use with quality process water e.g. dental surgeries.

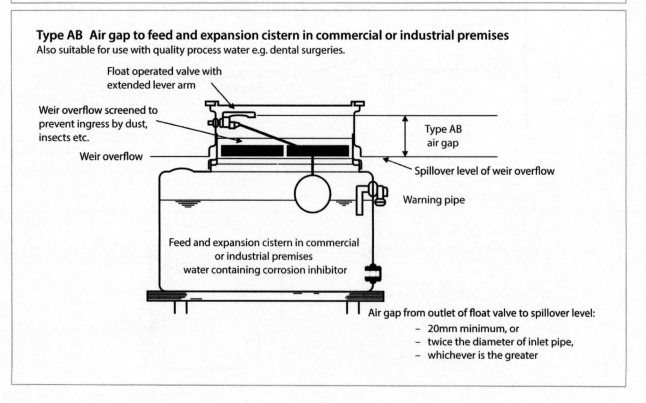

Float operated valve with extended lever arm

Weir overflow screened to prevent ingress by dust, insects etc.

Weir overflow

Type AB air gap

Spillover level of weir overflow

Warning pipe

Feed and expansion cistern in commercial or industrial premises
water containing corrosion inhibitor

Air gap from outlet of float valve to spillover level:
– 20mm minimum, or
– twice the diameter of inlet pipe,
– whichever is the greater

Type AC Air gap with submerged inlet and circular overflow discharging to a tundish

Means a non-mechanical backflow prevention arrangement of water fittings with a vented, but submerged, inlet; the air gap being measured vertically downwards from the lowest point of the air inlet to the critical water level.

Application of device: Backpressure – category 3
                                   Backsiphonage – category 3

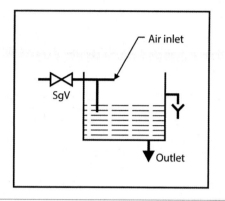

Type AD Air gap with injector and overflow
(often known as a 'jump jet')

Means a non-mechanical backflow prevention arrangement of water fittings with a horizontal injector and a physical air gap 20mm or twice the inlet diameter, whichever is the greater.

Application of device: Backpressure – category 5
                                   Backsiphonage – category 5

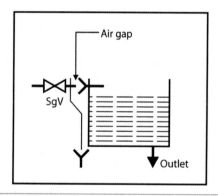

Type AF Air gap with circular overflow

Means a non-mechanical backflow prevention arrangement of water fittings with an air gap measured downwards from the lowest point of the discharge orifice, which discharges into the receptacle, to a critical level.

Application of device: Backpressure – category 4
                                   Backsiphonage – category 4

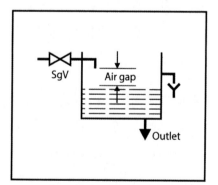

Type AG Air gap with minimum size circular overflow

(Satisfies requirements of Type B air gap to BS 6281 Part 2)

Means a non-mechanical backflow prevention arrangement of water fittings with an air gap; together with an overflow, the size of which is determined by measured or vacuum test.

Application of device: Backpressure – category 3
                                 Backsiphonage – category 3

Example of use: cold water storage cisterns for domestic purposes and interposed cisterns (see Type AUK1)

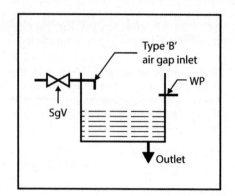

Application of Type AG air gap (Type B air gap to BS 6281 (Part 2)

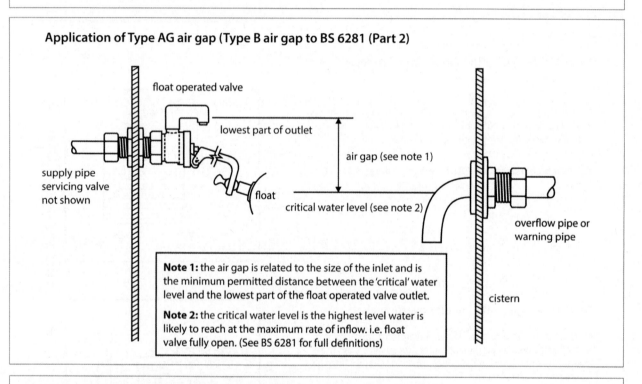

float operated valve

lowest part of outlet

air gap (see note 1)

supply pipe
servicing valve
not shown

float

critical water level (see note 2)

overflow pipe or
warning pipe

cistern

Note 1: the air gap is related to the size of the inlet and is the minimum permitted distance between the 'critical' water level and the lowest part of the float operated valve outlet.

Note 2: the critical water level is the highest level water is likely to reach at the maximum rate of inflow. i.e. float valve fully open. (See BS 6281 for full definitions)

Acceptable alternative to the Type B air gap

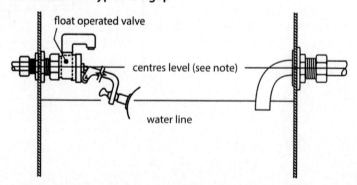

float operated valve

centres level (see note)

water line

Note: This arrangement is acceptable providing the cistern contains wholesome water and the installation complies with the requirements of paragraph 16 of Schedule 2 and the float operated valve is of the reducing flow type.

A float operated valve of the reducing flow type is one which gradually closes as the water level in the cistern rises. e.g. diaphragm valve to BS 1212 Part 2 or Part 3.

In this cistern, the critical water level is assumed to be level with the centre line of the flat operated valve.

Type AUK1 Air gap with interposed cistern

(incorporates Type B air gap to BS EN 14623:2005)

Means a non-mechanical backflow prevention arrangement consisting of a cistern with a Type AG overflow and an air gap; the spillover level of the receiving vessel (WC pan or other receptacle) being located not less than 300mm below the overflow pipe and not less than 15mm below the lowest level of the interposed cistern.

Application of device: Backpressure – category 3
Backsiphonage – category 5

Example of use: WCs and vessels containing high contamination risk fluids

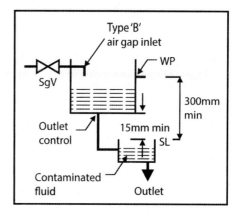

Application of Type AUK1 air gap to cattle drinking bowl with nozzle outlet below spillover level

Where outlet nozzle is below spillover level or is likely to be contaminated by animals mouth the bowl must be supplied from an interposed storage cistern from a pipe that only supplies similar appliances (see below)

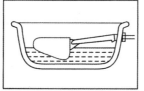

Interposed cistern to include type AG air gap

WC cistern acting as interposed cistern

WC cistern

15mm minimum

300mm minimum

WC pan

supply pipe or distributing pipe

Example of interposed cistern used to protect against contamination from cattle drinking bowls

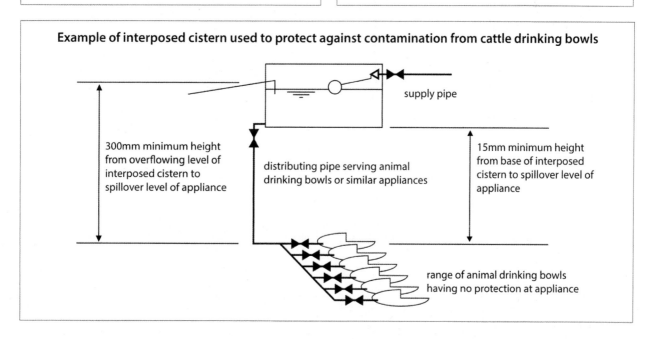

300mm minimum height from overflowing level of interposed cistern to spillover level of appliance

distributing pipe serving animal drinking bowls or similar appliances

supply pipe

15mm minimum height from base of interposed cistern to spillover level of appliance

range of animal drinking bowls having no protection at appliance

Type DC Pipe interrupter with permanent atmospheric vent

Means a non-mechanical backflow prevention device with a permanent unrestricted air inlet, the device being installed so that the flow of water is in a vertically downward direction.

Application of device: Backpressure – not acceptable
Backsiphonage – category 5

Example of use: Flush pipes to WCs and urinals

Except for urinals, this device must be fitted with the lowest point of the air aperture not less than 300mm above the free discharge point, or spillover level of an appliance, and have no valve, flow restrictor or tap on its outlet.

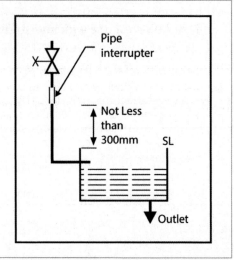

Application of Type DC pipe interrupter to a WC installation using pressure flushing valve

Permitted only in non-domestic premises

supply pipe or distributing pipe

modulely operated flushing valve min flow rate 1.21/s

pipe interrupter with permanent atmospheric vent

minimum 300mm

flush pipe

WC pan

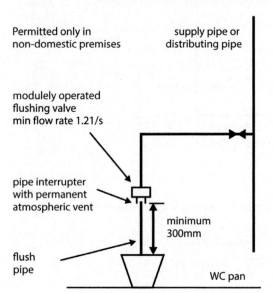

Interposed cistern to include Type AG air gap

supply pipe or distributing pipe

flushing valve, operated modulely or by automatic means

pipe interrupter with permanent atmospheric vent

150mm minimum

urinal bowl

Note: Lowest point of air inlet to pipe interrupter fitted at least 150mm above sparge pipe outlet to bowl

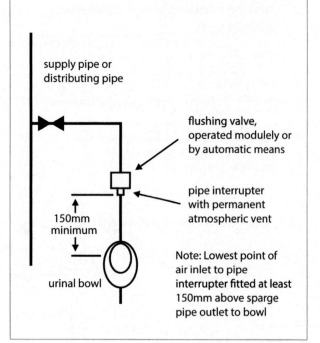

Now let's look at two more air gaps that have been mentioned but which perhaps need a little more attention. These are the Type AUK2 and Type AUK3 air gaps which are shown earlier in Table S6.1 on page 8.

These form the basis for all the air gaps and give vertical dimensions for air gaps across which water is not likely drawn should a backflow situation arise in the pipe supplying the tap or appliance served. The vertical dimension of the gap will depend on the diameter or size of the incoming pipe and on the severity of the risk.

For a full description you can refer back to page 8 or look in the glossary of terms (attached to this package). The following table will give a straightforward description of these air gaps.

Air gaps at taps, valves and fittings (including cisterns)

Situation	Nominal size of inlet of tap, valve or fitting	Vertical distance between tap or valve outlet and spillover level of receiving appliance (mm)
Domestic situations fluid category 2/3 device AUK2	Up to and including G½	20
	Over G½ and up to G¾	25
	Over G¾	70
Non-domestic situations fluid category 4/5 device AUK3	Any size inlet pipe	Minimum diameter 20mm, or twice the diameter of the inlet pipe, whichever is the greatest

Note: AUK2 and AUK3 devices are suitable for protection against backsiphonage risk only.

Mechanically operated backflow prevention devices

Mechanically operated devices will all need to be periodically inspected and maintained. Any mechanical device can go wrong, so the maintenance of these devices is important. This is especially the case for the RPZ valve (below) where regular inspection and testing is a requirement for its use, testing shall be carried out at least annually or at more frequent intervals as specified by the water supplier.

Type BA Verifiable backflow preventer with reduced pressure zone
(commonly known as an 'RPZ valve assembly')

Means a verifiable mechanical backflow prevention device consisting of an arrangement of water fittings with three pressure zones with differential obturators and that will operate when potential backflow conditions occur or there is a malfunction of the valve.

A Type AA air gap should be provided between the relief outlet port and the top of the allied tundish.

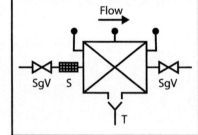

Application of device: Backpressure – category 4
Backsiphonage – category 4

Example of use: For protection against installations or appliances of fluid risk category 4. e.g. Fire sprinkler systems with anti-freeze additives in commercial or industrial premises.

Approved Installation Method (AIM-08-01) for BA Device – a Verifiable Backflow Preventer with Reduced Pressure Zone – that includes changes within Approved Contractor Schemes T&Cs which require Approved Contractor's to give prior notice to the relevant water undertaker, for the installation of all RPZ valves. For further information see https://www.waterregsuk.co.uk/guidance/publications/approved-installation/

Installation, commissioning and testing of an RPZ valve must only be carried out be an accredited tester approved by the Water Supplier as being competent to test. On completion of a test, a 'test report' certificate must be completed by the tester and copies submitted to the person responsible for the device and to the Water Undertaker within 10 working days.

A test record card is recommended to be left on or adjacent to the RPZ valve.

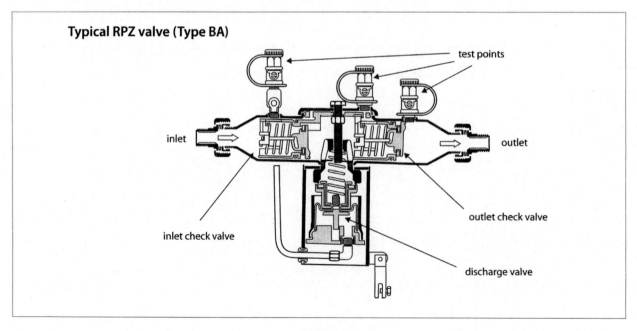

Typical RPZ valve (Type BA)

Example of RPZ valve installation

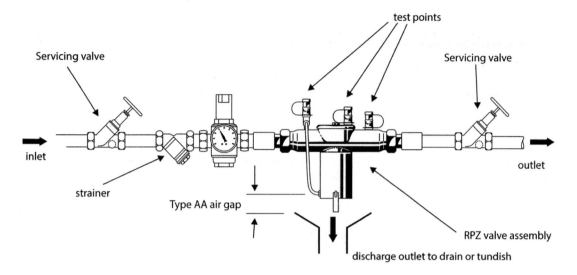

Installation notes:

- Install in acceptable position, free from flooding and from effects of frost, not above electrical equipment, and in a secure and lockable cabinet.

- Assembly to be installed above ground in horizontal position (unless otherwise approved) and at a height above the ground that will permit effective maintenance.

- Clearance should be allowed for test equipment to be fitted and to permit maintenance,
minimum height from floor or cabinet	– 300mm
maximum height from the ground	– 1500mm
minimum space at rear	– 50mm sizes DN15-DN50) and
minimum space at rear	– 100mm (sizes DN65-DN250) and
minimum space in front	– 200mm

- Adequate drainage should be provided, and a Type AA air gap arranged between relief valve outlet and tundish/drain.

- No drinking water supply to be drawn off downstream of an RPZ assembly.

- RPZ valve assemblies will cause a drop in pressure and may not be suitable for use on low pressure supplies.

- Assemblies to be flushed out and disinfected after installation and before use.

- Assembly to be checked following installation to ensure relief valves function correctly.

- Assembly to be 'site tested' before use and at regular intervals after installation by an accredited tester and test certificate issued after each test.

- Records should be kept of all installation and maintenance procedures. Copies to be retained by both installer and the customer.

- Accredited testers will need to be approved with the Water Supplier.

- Water Supplier to be notified and approval given before RPZ valve is installed, and should be notified of any maintenance tests carried out.

- For more comprehensive information refer to:
  - **Information and Guidance Note No 9-03-02** issued by the Water Regulations Advisory Scheme, or alternatively the Water Supplier should be able to advise.

Type CA Non-verifiable disconnector with different pressure zones

Means a non-verifiable mechanical backflow prevention device which provides disconnection by venting the intermediate pressure zone of the device to the atmosphere when the difference of pressure between the intermediate zone and the upstream zone is not greater than 10% of the upstream pressure.

A Type AA air gap should be provided between the relief outlet port and the top of the allied tundish.

Application of device: Backpressure – category 3
Backsiphonage – category 3

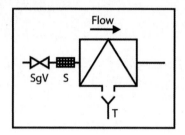

Typical Non-verifiable disconnector with different pressure zones

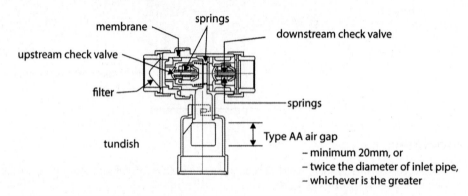

membrane

springs

downstream check valve

upstream check valve

filter

springs

tundish

Type AA air gap
– minimum 20mm, or
– twice the diameter of inlet pipe,
– whichever is the greater

Type DA Anti-vacuum valve (or vacuum breaker)

Means a mechanical backflow prevention device with an air inlet which is closed when water in the device is at or above atmospheric pressure but which opens to admit air if a vacuum occurs at the inlet to the device.

The device must be fitted on a Type A upstand so that the outlet is not less than 300mm above the free discharge point, or spillover level, and have no valve or restriction on its outlet.

Application of device: Backpressure – not acceptable
Backsiphonage – category 3

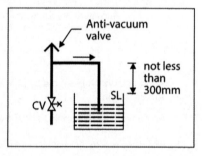

Typical in-line anti-vacuum valve

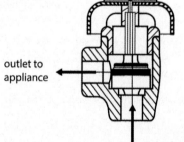

outlet to appliance

Under backsiphonage conditions, valve will allow downstream (inlet) pipe to empty without drawing water from outlet pipe or appliance

No means of closure may be fitted downstream between outlet of valve and point of delivery to appliance

Outlet to be at least 300mm above free discharge point or spillover level of appliance

inlet from supply pipe or distributing pipe

Type DB   Pipe interrupter with vent and moving element

Means a mechanical backflow protection device with an air inlet closed by a moving element when the device is in normal use but which opens and admits air if the water pressure upstream of the device falls to atmospheric pressure, the device being installed so that the flow of water is in a vertical downward direction.

The device is to be fitted with the lowest point of the air aperture not less than 300mm above the free discharge point, or spillover level, and have no valve, flow restrictor or tap on its outlet.

Application of device: Backpressure – not acceptable
                                    Backsiphonage – category 4

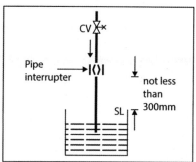

Example of use: Mini-irrigation system or porous hose for garden watering in houses

---

Typical Type DB Pipe interrupter

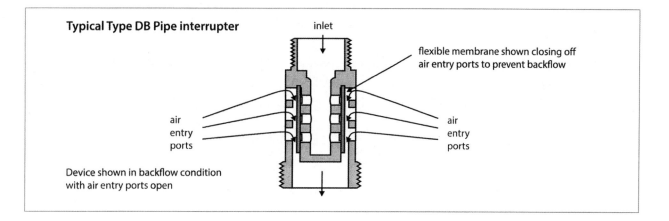

inlet

flexible membrane shown closing off air entry ports to prevent backflow

air entry ports

air entry ports

Device shown in backflow condition with air entry ports open

Type DUK1   Anti-vacuum valve combined with verifiable check valve

Means a mechanical backflow prevention device comprising an anti-vacuum valve with a single check valve located upstream.

The device must be fitted on a Type B upstand so that the outlet of the device is not less than 300mm above the free discharge point, or spillover level, and have no valve, flow restrictor or tap on its outlet.

Application of device: Backpressure – category 2
                                    Backsiphonage – category 3

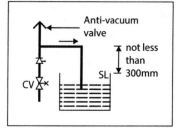

Typical in-line anti-vacuum valve

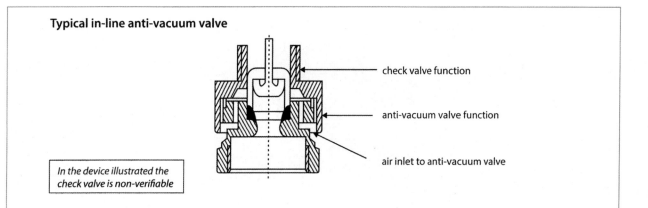

check valve function

anti-vacuum valve function

air inlet to anti-vacuum valve

In the device illustrated the check valve is non-verifiable

Type EA Verifiable single check valve

Means a verifiable mechanical backflow prevention device which will permit water to flow from upstream to downstream but not in the reverse direction.

Application of device: Backpressure – category 2
                                            Backsiphonage – category 2

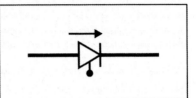

Typical Type EA Verifiable single check

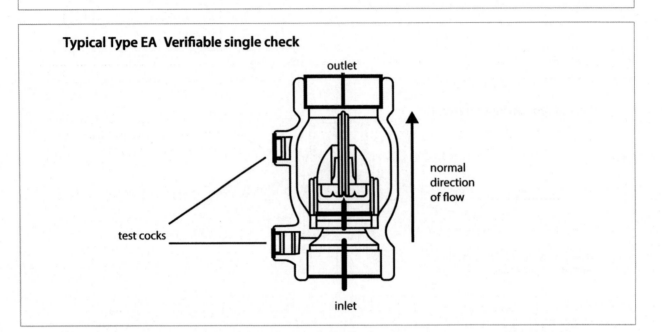

Type EB Non-verifiable single check valve

Means a non-verifiable mechanical backflow prevention device which will permit water to flow from upstream to downstream but not in the reverse direction.

Application of device: Backpressure – category 2
                                            Backsiphonage – category 2

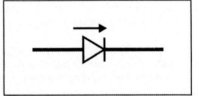

Typical non-verifiable single check valve

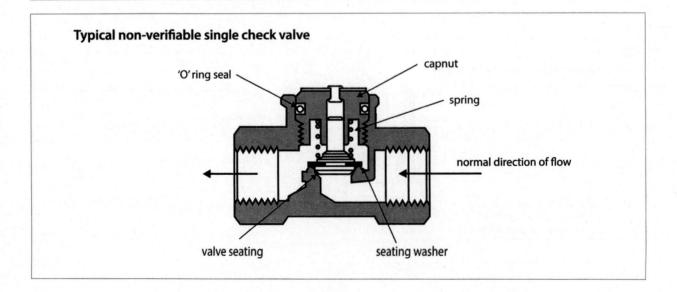

Type EC Verifiable double check valve

Means a verifiable mechanical backflow prevention device consisting of two verifiable single check valves in series, which will permit water to flow from upstream to downstream but not in the reverse direction.

Application of device: Backpressure – category 3
Backsiphonage – category 3

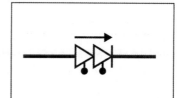

Typical verifiable double check valve

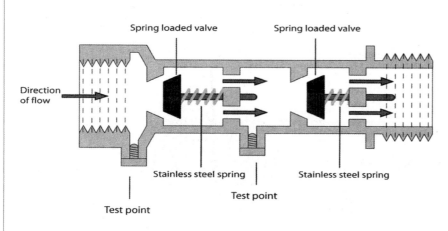

Spring loaded valve

Spring loaded valve

Direction of flow

Stainless steel spring

Stainless steel spring

Test point

Test point

Type EC double check valves are used in exactly the same way as single check valves. They are basically two single check valves in tandem with one another. Type EC valves are verifiable this means they have a test point where they can be checked for the operation of the valve. Used for fluid category 1 protection against fluid category 3 risk i.e. outside tap installations and central heating temporary filling loops (in conjunction with an air gap)

Type ED Non-verifiable double check valve

Means a non-verifiable mechanical backflow prevention device consisting of two non-verifiable single check valves in series, which will permit water to flow from upstream to downstream but not in the reverse direction.

Application of device: Backpressure – category 3
Backsiphonage – category 3

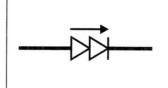

Typical non-verifiable double check valve

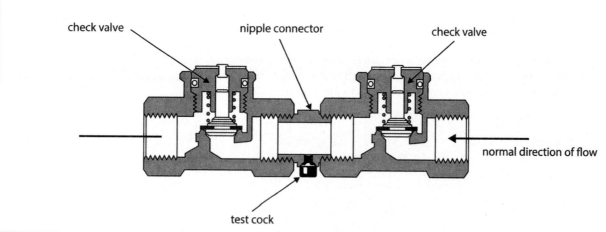

check valve

nipple connector

check valve

normal direction of flow

test cock

Type HA Hose union backflow preventer

Only permitted for use on existing hose union taps in house installations.

Means a mechanical backflow prevention device for fitting to the outlet of a hose union tap and consisting of a single check valve with air inlets that open if the flow of water ceases.

Not to be used in new installations. Only permitted outside houses for replacement of existing hose union taps that do not incorporate any backflow device.

Application of device: Backpressure – category 2
                              Backsiphonage – category 3

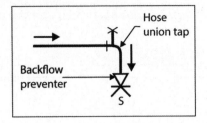

Type HA Hose union backflow preventer

normal direction of flow

Device shown in backflow condition

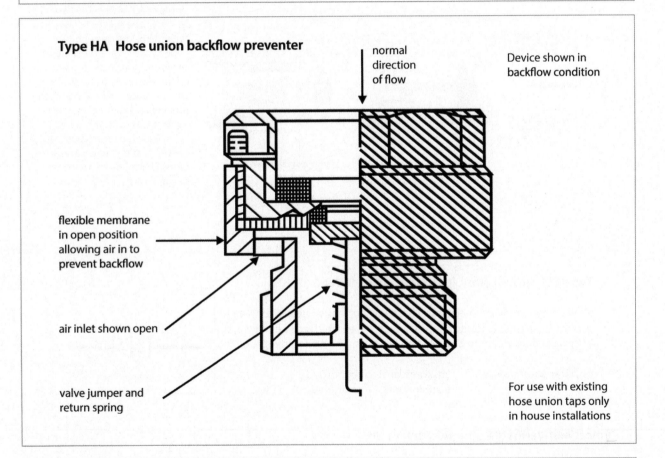

flexible membrane in open position allowing air in to prevent backflow

air inlet shown open

valve jumper and return spring

For use with existing hose union taps only in house installations

Type HC Diverter with automatic return

Means a mechanical backflow device used in bath/shower combination tap assemblies which automatically returns the bath outlet open to atmosphere if a vacuum occurs at the inlet to the device.

Integral with some domestic appliances only.

Application of device: Backpressure – not applicable
                              Backsiphonage – category 3

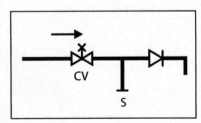

Type HUK1 Hose union tap which incorporates verifiable double check valve

Means a hose union tap in which a double check valve has been incorporated into the inlet or the outlet of the tap.

Not for use with new installations. Only permitted outside houses for replacement of existing hose union taps that do not incorporate any backflow device.

Application of device: Backpressure – category 3

                                 Backsiphonage – category 3

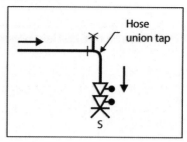

Example of Type HUK1 Hose union tap – screwdown type

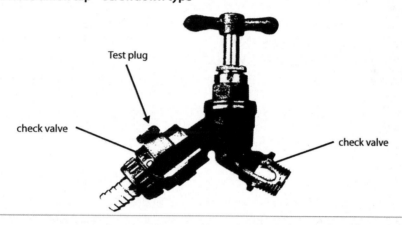

Type LA Pressurised air inlet valve

Means an anti-vacuum valve or vacuum breaker, similar to Type DA but suitable for conditions where the water pressure at the outlet of the device under normal conditions of use is greater than atmospheric.

Use is limited to locations where operational waste is acceptable, e.g. in gardens or similar.

Application of device: Backpressure – not applicable

                                 Backsiphonage – category 2

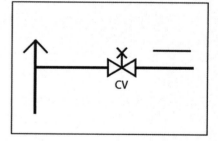

Type LB Pressurised air inlet valve with a check valve downstream

Means a mechanical backflow prevention device comprising a Type LA anti-vacuum valve with a single check valve located downstream.

Use is limited to locations where operational waste is acceptable, e.g. in gardens or similar.

Application of device: Backpressure – category 2

                                 Backsiphonage – category 3

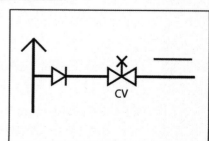

Self-assessment questions (2)

1. The following table gives a list of non-mechanically operated backflow prevention devices. Using Table S6.1 on page 8, complete the table below to show the minimum category of risk for which each backflow device is suited.

Type of device	Description of arrangement or device	Minimum category of backpressure risk to which it is suited	Minimum category of backsiphonage risk to which it is suited
Type AA	Air gap with unrestricted discharge (formerly Type A air gap)	_____	_____
Type AB	Air gap with weir overflow	_____	_____
Type AG	Air gap with minimum size circular overflow (satisfies requirements of Type B air gap to BS 6281 Part 2)	_____	_____
Type AUK1	Air gap with interposed cistern	_____	_____
Type DC	Pipe interrupter with permanent atmospheric vent	_____	_____

2. The following table gives a list of mechanically operated backflow prevention devices. Using Table S6.2 on page 9 complete the table below to show the minimum category of risk for which each backflow device is suited.

Type of device	Description of arrangement or device	Minimum category of backpressure risk to which it is suited	Minimum category of backsiphonage risk to which it is suited
Type BA	Verifiable backflow preventer with reduced pressure zones	_____	_____
Type DB	Pipe interrupter with atmospheric vent and moving element	_____	_____
Type EA	Verifiable single check valve	_____	_____
Type EB	Non-verifiable single check valve	_____	_____
Type EC	Verifiable double check valve	_____	_____
Type ED	Non-verifiable double check valve	_____	_____
Type HUK1	Hose union tap which incorporates a double check valve	_____	_____

3. A Type AA air gap should have an unrestricted discharge delivering vertically downwards and its dimension depends on the size of the inlet to the vessel served.

Complete the following rules relating to a Type AA air gap.

The Type AA air gap at point of use should be:

minimum – _____, *or*

– _____,

– *whichever is the greater.*

4. Complete the following table of air gaps at taps, valves and fittings.

Situation	Nominal size of inlet of tap, valve or fitting	Vertical distance between tap or valve outlet and spillover level of receiving appliance (mm)
Domestic situations fluid category 2/3 device AUK2	Up to and including G½	_____
	Over G½ and up to G¾	_____
	Over G¾	_____
Non-domestic situations fluid category 4/5 device AUK3	any size inlet pipe	minimum diameter _____ mm, or twice the diameter of the inlet pipe, whichever is the greatest
Note: AUK2 and AUK3 devices are suitable for protection against backsiphonage risk only		

Check your answers on pages 69 and 70.

So far we've looked at the various backflow devices, but:

What about their practical application to everyday installations?

Well! You have seen a few examples but lets look at some more.

Application of backflow devices

In general backflow can be prevented by good system design, combined with the use of backflow devices or arrangements, chosen to suit the category of risk for which they are designed.

The Regulator's Specification on the Prevention of Backflow gives a number of examples for a range of applications:

WCs and urinals

WC pans and urinals are considered to be a fluid category 5 risk and present a serious contamination risk no matter what their situation, whether in a house or in industrial or commercial premises.

There are two suitable types of backflow device for these:

1. an **interposed cistern (Type AUK1),** which means a siphonic or non-siphonic flushing cistern, which **may be used in premises of any type;**

2. a **pipe interrupter with permanent atmospheric vent (Type DC),** fitted to the outlet of a modulely operated pressure flushing valve, **may be connected to a supply pipe or distributing pipe (but not in a house).** There should be no other obstruction between the outlet of the pipe interrupter and the flush pipe connection to the appliance.

   The pipe interrupter **should be positioned** so that its vent outlet is at least:

   - **300mm above the overflowing level of the WC pan, or**
   - **150mm above the urinal bowl being served.**

WC pan supplied from interposed cistern (flushing cistern)

WC cistern acting as interposed cistern

Interposed cistern to include Type AG air gap

Cistern overflowing level to be at least 300mm above spillover level of WC pan

Base of cistern to be at least 15mm above spillover level of WC pan

WC cistern

15mm minimum

300mm minimum

WC pan

supply pipe or distributing pipe

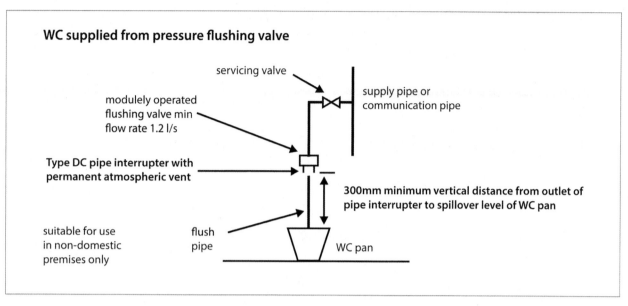

WC supplied from pressure flushing valve

servicing valve

supply pipe or communication pipe

modulely operated flushing valve min flow rate 1.2 l/s

Type DC pipe interrupter with permanent atmospheric vent

300mm minimum vertical distance from outlet of pipe interrupter to spillover level of WC pan

suitable for use in non-domestic premises only

flush pipe

WC pan

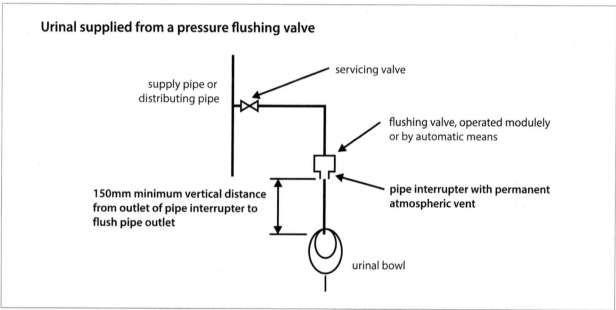

Urinal supplied from a pressure flushing valve

supply pipe or distributing pipe

servicing valve

flushing valve, operated modulely or by automatic means

pipe interrupter with permanent atmospheric vent

150mm minimum vertical distance from outlet of pipe interrupter to flush pipe outlet

urinal bowl

Are bidets as great a risk as WCs?

Yes they are! **Bidets are classified as fluid risk category 5 appliances** and as such present a serious health hazard to users. So! We'll look at these next.

Bidets and WCs adapted as bidets.

For the purposes of the regulations, bidets (and WCs adapted as bidets) can be divided into two groups, each of which has different protection applied:

1. **Over rim types,** those that are supplied from a tap at the back edge of the appliance; and

2. **Ascending spray** or submerged inlet types, those that have a water delivery spray jet which is situated below the rim or over spill level of the bowl. Included in this group are those bidets that are supplied from a tap but use a flexible spray attachment.

We shall refer to bidets in the following notes, but anything we say can equally be applied to WCs adapted as bidets.

Over-rim type bidets are supplied from taps in a similar way to a wash basin. The important thing to remember here is that **an air gap (tap gap) is needed between the tap outlet and the spillover level of the appliance**. Despite the fact that a bidet is a fluid category 5 contamination risk, guidance to the regulations recommends that over-rim type bidets have a Type AUK2 air gap, this protects normally against a fluid category contamination risk. This is illustrated in the following diagram.

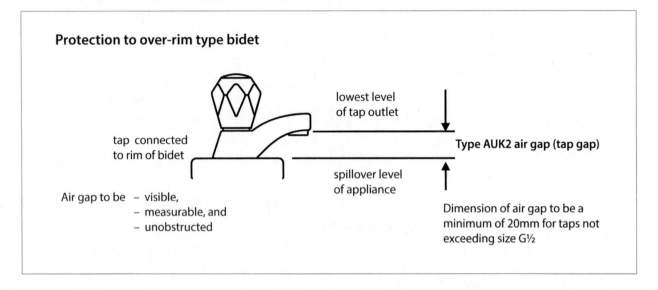

Protection to over-rim type bidet

tap connected
to rim of bidet

lowest level
of tap outlet

Type AUK2 air gap (tap gap)

spillover level
of appliance

Air gap to be – visible,
– measurable, and
– unobstructed

Dimension of air gap to be a
minimum of 20mm for taps not
exceeding size G½

Providing that appliances and taps comply to relevant British or European Standards, or are listed in the WRAS Fittings and Materials Directory, the application of the regulations to over-rim type bidets is not a big problem. Manufacturers should sort this out for us. All we need to do is measure the air gap and compare it to the size of the inlet pipe. This will be particularly important with appliances and taps supplied from abroad.

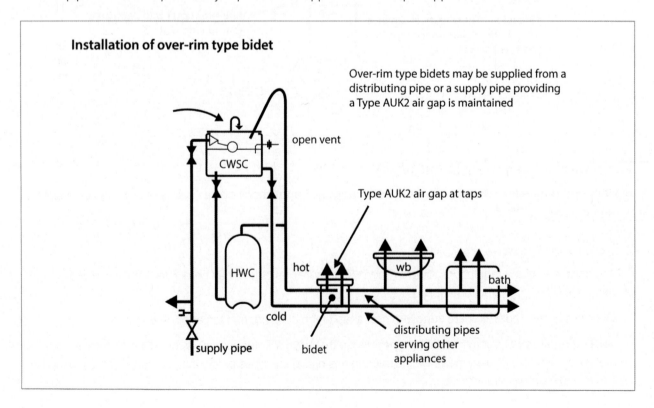

Installation of over-rim type bidet

Over-rim type bidets may be supplied from a
distributing pipe or a supply pipe providing
a Type AUK2 air gap is maintained

open vent

CWSC

Type AUK2 air gap at taps

HWC   hot   wb   bath

cold

supply pipe   bidet

distributing pipes
serving other
appliances

Ascending spray type bidets

When making connections the following points should be borne in mind.

Ascending spray type bidets, or over rim bidet with a flexible hose attachment, are not permitted to be connected directly from the supply pipe and must therefore be supplied from a storage (break) cistern.

Both **hot and cold connections** should be taken from **independent** dedicated **distributing pipes that do not supply other appliances;** except in the following cases:

An ascending spray type bidet may be supplied by a common distributing pipe but only when:

(a) the common distributing pipe **serves only the bidet and a WC or urinal flushing cistern**

(b) the bidet is the **lowest fitting served** and there is no likelihood of other fittings being connected at a lower level at a later date, and the connection to the common distributing pipe is not less than 300mm above the spillover level of the bowl, or

(c) in the case of an over-rim bidet with a flexible spray connection, not less than 300mm above the spillover level of any appliance that the spray outlet might reach.

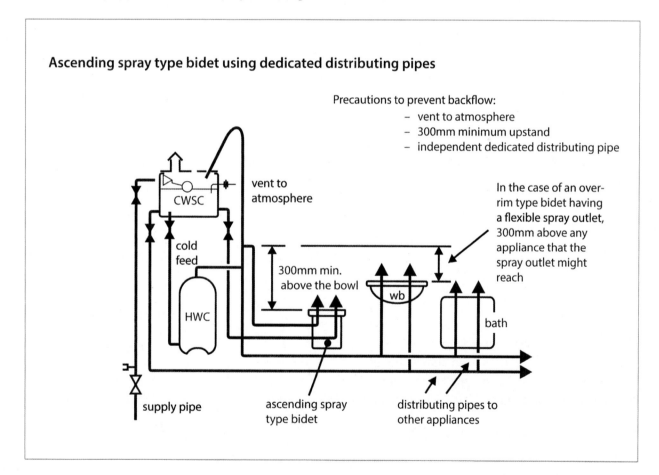

Ascending spray type bidet using dedicated distributing pipes

Precautions to prevent backflow:
- vent to atmosphere
- 300mm minimum upstand
- independent dedicated distributing pipe

In the case of an over-rim type bidet having a flexible spray outlet, 300mm above any appliance that the spray outlet might reach

CWSC

vent to atmosphere

cold feed

300mm min. above the bowl

wb

HWC

bath

supply pipe

ascending spray type bidet

distributing pipes to other appliances

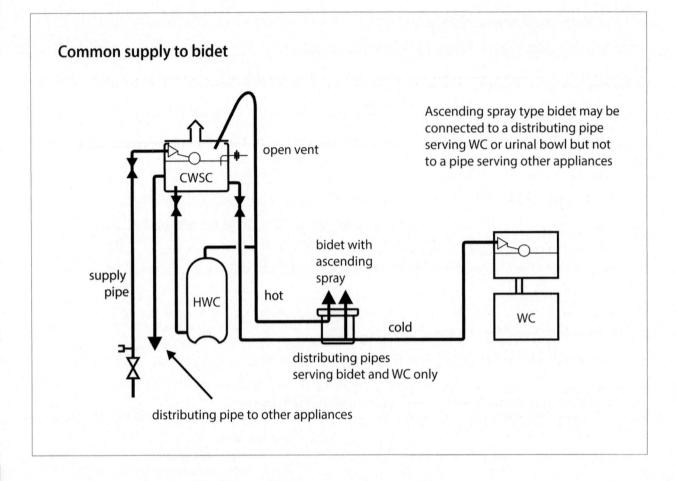

Common supply to bidet

open vent

CWSC

Ascending spray type bidet may be connected to a distributing pipe serving WC or urinal bowl but not to a pipe serving other appliances

supply pipe

HWC

hot

bidet with ascending spray

cold

WC

distributing pipes serving bidet and WC only

distributing pipe to other appliances

What about connections to other types of appliance?

Let's look at some!

Taps and shower outlets to sanitary appliances

Using taps without the correct backflow protection puts users at risk of contaminating their drinking water.

Single outlet taps, combination taps, pull out spouts and pull down taps fixed shower heads should discharge above the appliance and terminate with an air gap.

In domestic premises a Type AUK 2 tap gap should be used.

Tap gaps for taps and shower outlets

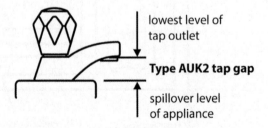

lowest level of tap outlet

Type AUK2 tap gap

spillover level of appliance

Type AUK2 air gap for up to category 3 risk against backsiphonage only, e.g. baths, basins and showers in domestic premises

The Regulations and Byelaws require the preventing of backflow from a sink through the tap.

Sinks in both domestic and non-domestic situations are considered to be a fluid category 5 backflow risk and as such the minimum protection at the sink is the Type AUK3 air gap (tap gap). However, this is not generally a problem as sinks need additional space for access to work and for the filling of buckets and other appliances.

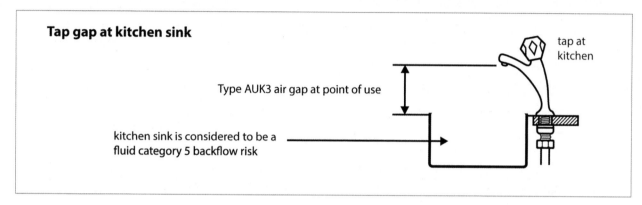

Tap gap at kitchen sink

tap at kitchen

Type AUK3 air gap at point of use

kitchen sink is considered to be a fluid category 5 backflow risk

Where innovative designs for domestic kitchen sink taps include taps with pull-out spouts, taps which fold into the sink below the worktop level and the introduction to the domestic market of 'pull-down' taps similar to those used in commercial kitchens can cause a **Risk to health.**

These types of taps risk drawing contaminated water back into the plumbing system. **Pull-out spouts** risk being left with the discharge point lying in the sink. **Pull-down taps** risk having the flexible spout trapped below the spillover level and fold-down taps are designed to be stored with their outlets below the spillover level.

Taps should be designed and installed so that their spouts are clear of the water in a sink.

Keeping the law: In the UK the installer of plumbing fittings which will be supplied from the public supply has a legal duty to ensure that an installation meets the requirements of the Water Supply (Water Fittings) Regulations or Byelaws in Scotland. The user, owner or occupier also have legal duties to ensure these requirements are met.

Traditional tap gap arrangement for sinks – domestic or non-domestic

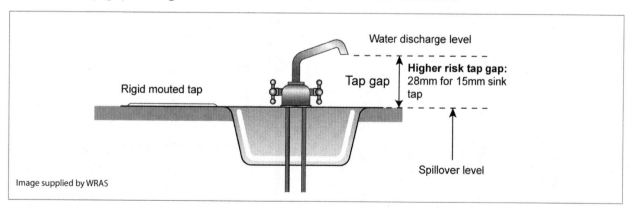

Water discharge level

Rigid mouted tap

Tap gap

Higher risk tap gap: 28mm for 15mm sink tap

Spillover level

Image supplied by WRAS

Pull-out spouts and Pull-down taps: It is a legal requirement for all plumbing designs of new homes to be notified to the local water company, in advance of starting work, to obtain their consent.

Where new homeowners request changes to the original design such as installing these types of taps, changes must be re-notified and consent granted. Where refurbishing existing homes it's wise to consult the local water company and obtain their consent, given the additional risks of these installations.

Taps used in high risk installations, such as kitchen sinks, are a fluid category 5 risk. This requires a physical air break to provide the category 5 protection needed as provided by the 'Higher risk tap gap'. With most **fold-down**, **pull-down** and **pull-out** taps not having a fixed outlet, cannot themselves provide this type of backflow protection and must not be connected directly to the drinking water supply. Therefore additional backflow protection must be provided before the tap on the hot and cold supplies.

The type of protection will depend on the manufacturer's requirements for the tap, as some require minimum levels of flow or water pressure to operate.

Appropriate fluid category 5 protection built in

Gravity fed storage: the usual method is for the water supplies to be fed via storage cisterns (or as they are sometimes referred to 'break tanks') which provide fluid category five backflow protection.

Two suitable arrangements can been seen in the images below, one of which uses an elevated cistern with an air gap to supply the hot and cold feeds to the taps. The other using boost pumps.

The first method may not suit all types of taps as the water pressure may be low due to being fed via the cistern, check with manufacture's specifications prior to installation.

The second method will require the water supplier consent if the flow rate of the pumps exceeds 12 litres per minute.

Method for using a storage cistern to provide the hot and cold water

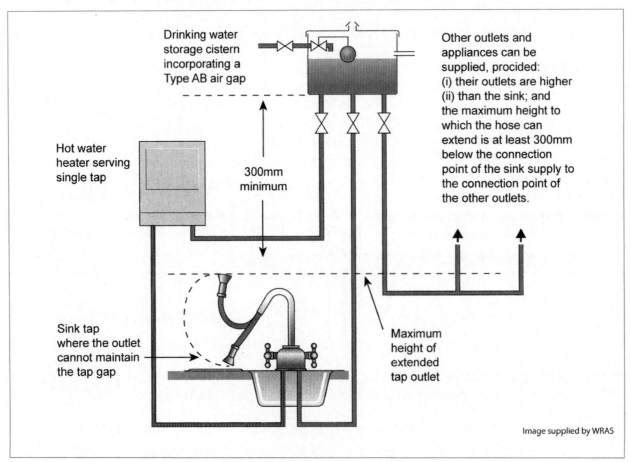

Image supplied by WRAS

Method for using a storage cistern with fluid category 5 air gap and booster pumps to provide backflow protection, (note the cistern cannot supply outlets other than the sink taps due to backflow issues involving the pumps)

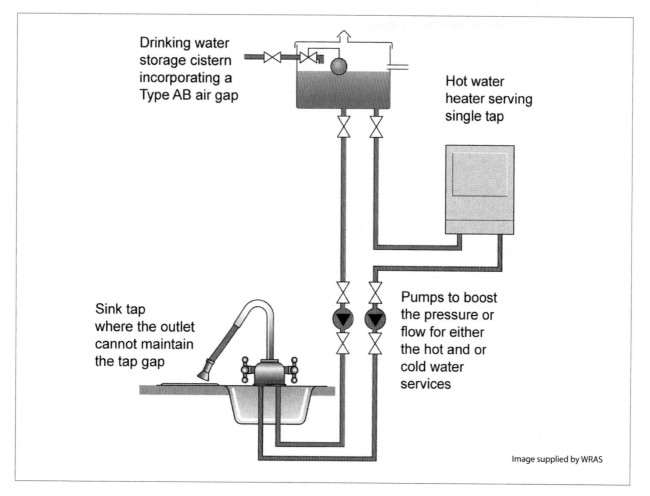

Drinking water storage cistern incorporating a Type AB air gap

Hot water heater serving single tap

Sink tap where the outlet cannot maintain the tap gap

Pumps to boost the pressure or flow for either the hot and or cold water services

Image supplied by WRAS

Where appliances such as **baths and wash basins in domestic premises have submerged inlets they are considered to be a category 3** risk and both hot and cold inlets should be supplied through Type EC, or ED double check valves.

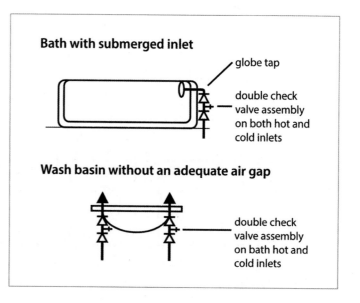

Bath with submerged inlet

globe tap

double check valve assembly on both hot and cold inlets

Wash basin without an adequate air gap

double check valve assembly on bath hot and cold inlets

Washing machines and dishwashers

Household machines should have backflow protection to fluid level 3 built in during manufacture. Before installing an appliance the Water Fittings and Materials Directory should be consulted. They will be listed if they have been approved under the Water Regulations Advisory Scheme.

Where a connection hose is not approved, a check valve is required to prevent backflow from the hose.

Commercial machines such as those used in laundromats or similar premises are a category 4 risk, whilst clothes washing plant or equipment in health care establishments are fluid category 5.

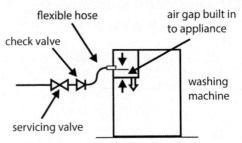

Domestic washing machine

flexible hose

air gap built in to appliance

check valve

servicing valve

washing machine

If the hosepipe to the washing machine is not listed in the Materials and Fittings Directory a check value must be fitted near the connection to the hose.

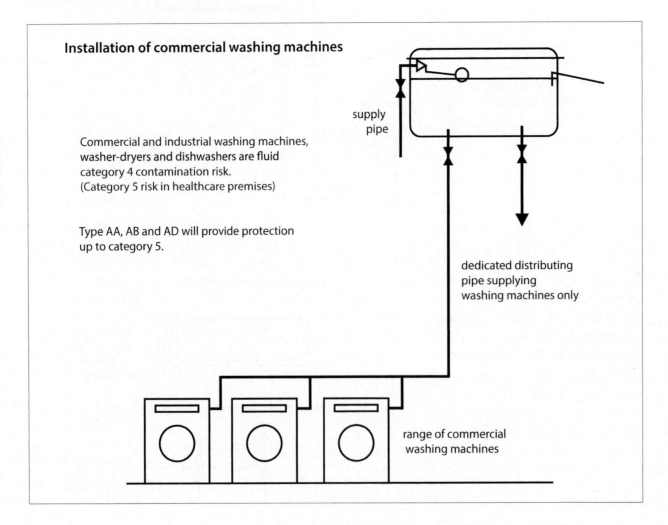

Installation of commercial washing machines

Commercial and industrial washing machines, washer-dryers and dishwashers are fluid category 4 contamination risk.
(Category 5 risk in healthcare premises)

Type AA, AB and AD will provide protection up to category 5.

supply pipe

dedicated distributing pipe supplying washing machines only

range of commercial washing machines

Drinking water fountains

These should be designed so that there is a minimum **25mm air gap** between the water delivery nozzle and the spillover level of the bowl.

Additionally **the nozzle should be screened or shrouded** to prevent mouth contact.

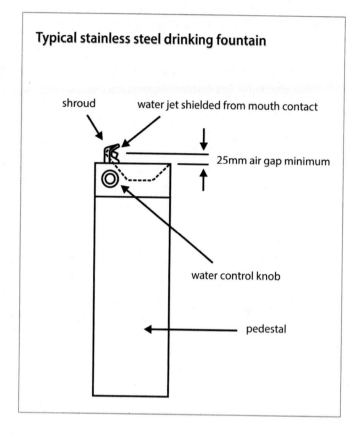

Typical stainless steel drinking fountain

shroud

water jet shielded from mouth contact

25mm air gap minimum

water control knob

pedestal

Self-assessment questions (3)

1. To make backflow precautions effective in flushing cisterns and devices, there are rules applied to the installation of them. Complete the following diagrams to show correct minimum dimensions for effective backflow prevention.

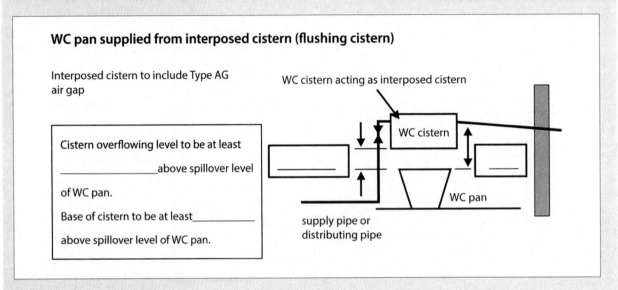

WC pan supplied from interposed cistern (flushing cistern)

Interposed cistern to include Type AG air gap

WC cistern acting as interposed cistern

Cistern overflowing level to be at least

_____above spillover level of WC pan.

Base of cistern to be at least_____

above spillover level of WC pan.

WC cistern

WC pan

supply pipe or distributing pipe

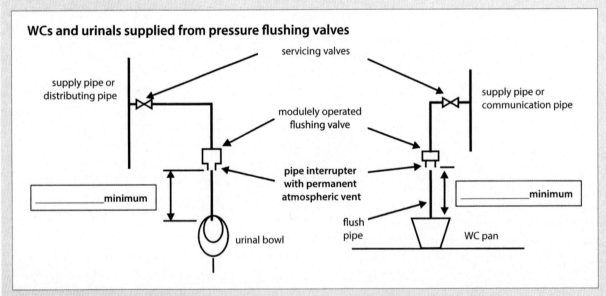

WCs and urinals supplied from pressure flushing valves

servicing valves

supply pipe or distributing pipe

supply pipe or communication pipe

modulely operated flushing valve

pipe interrupter with permanent atmospheric vent

_____minimum

_____minimum

flush pipe

urinal bowl

WC pan

2. Give the correct backflow protection device to suit an over-rim type bidet.

   a) in domestic premises _____

   b) in a hospital _____

3. Are ascending spray type bidets permitted to be connected directly from the supply pipe?
   Answer: Yes ☐   No ☐

4. Complete the following diagram to show an acceptable method of connecting an ascending spray type bidet. Also complete the labelling to give precautions that should be applied to the installation.

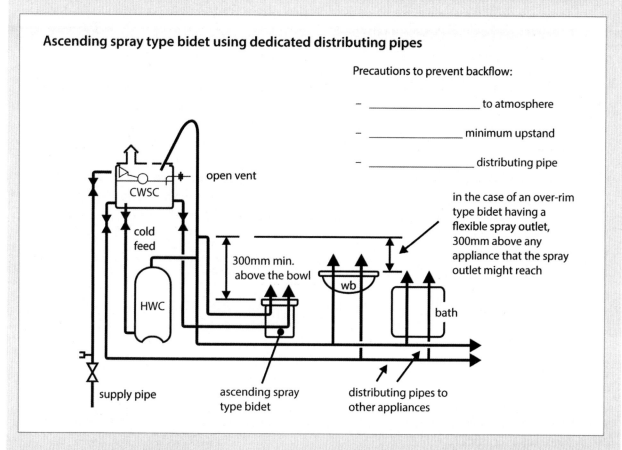

Ascending spray type bidet using dedicated distributing pipes

Precautions to prevent backflow:

– _____ to atmosphere

– _____ minimum upstand

– _____ distributing pipe

in the case of an over-rim type bidet having a flexible spray outlet, 300mm above any appliance that the spray outlet might reach

open vent

CWSC

cold feed

300mm min. above the bowl

HWC

wb

bath

supply pipe

ascending spray type bidet

distributing pipes to other appliances

5. Taps and showers need different levels of backflow protection depending on their risk assessment. In each of the following cases, state the fluid risk category and the minimum protection needed.

Application of backflow protection to taps and shower fittings	Fluid risk category	Minimum suitable backflow protection needed
Single outlet and combination taps, and shower heads to wash basins and baths in domestic premises		
Single outlet and combination taps, to sinks in domestic and commercial premises		
Single outlet and combination taps to baths and basins in hospital premises		

6. Baths and basins in domestic premises are a fluid category 3 risk whilst sinks are category 5.

   a) State the usual backflow protection given to

      i)   a domestic bath or basin_____

      ii)   a sink in domestic premises_____

   b) State suitable backflow protection for a basin, bath or shower where the water supply outlet from
      tap or shower head is below, or could be placed below the spillover level of the appliance.

      _____

7. Washing machines and dishwashers in domestic premises should have protection built in during
   manufacture. However, if you come across a domestic machine with connection hoses that are not
   WRAS approved, what protection should be given to the supply or distributing pipe.

   _____

   _____

8. Commercial washing machines are a more serious risk than domestic ones. Give a suitable backflow
   protection device for a range of commercial washing machines in a laundromat.

   _____

9. Complete the following diagram to show two important contamination prevention requirements for a
   drinking water fountain.

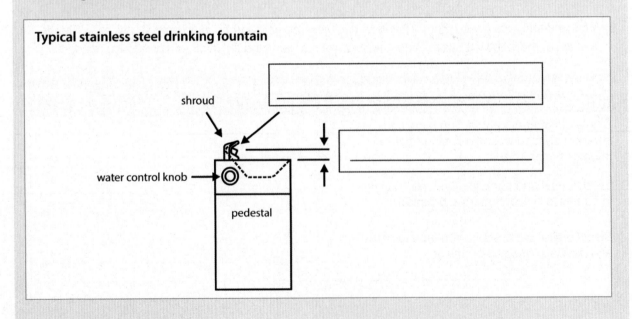

Typical stainless steel drinking fountain

shroud

water control knob

pedestal

Check your answers on pages 71, 72 and 73.

Are we going to look at outside taps?

Yes! This is becoming quite complicated because there are so many different situations involving a variety of fluid risk categories, and of course the risk category determines which protection device should be applied.

Outside taps and garden supplies

Backflow protection for hose taps depends on the level of risk for the individual application. Let's look at a few rules:

Hosepipes held in the hand for garden and other uses **should be fitted with a self-closing mechanism at the hose outlet**. (G15.17)

In a house situation:

- **any garden tap to which a hose connection can be made**
- **should be fitted with a double checkvalve**
- **positioned inside a building** where it will not be subjected to frost damage. (G15.20).

This will give adequate protection if the hose is held in the hand or where the hose outlet is fixed so that it has a permanent air gap.

The **double check valve** is also considered **sufficient protection for hand held hoses used for spraying fertilisers or domestic detergents in house garden situations**.

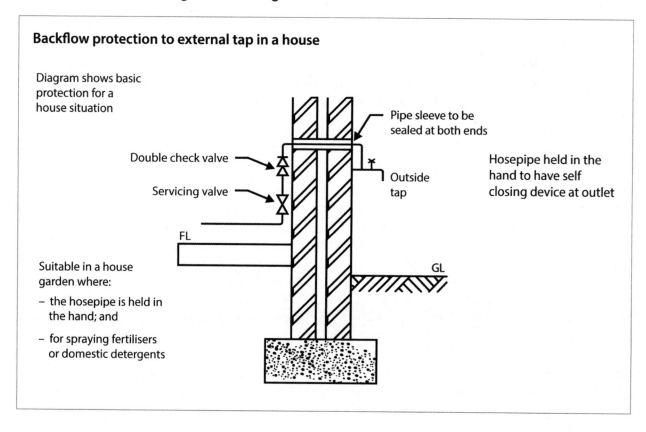

Backflow protection to external tap in a house

Diagram shows basic protection for a house situation

Double check valve

Servicing valve

FL

Pipe sleeve to be sealed at both ends

Outside tap

GL

Hosepipe held in the hand to have self closing device at outlet

Suitable in a house garden where:

- the hosepipe is held in the hand; and
- for spraying fertilisers or domestic detergents

For the spraying of insecticides (fluid category 5) suitable protection would be **a pipe interrupter with moving element (Type DB) fitted at the connection of the hose and at least 300mm above the highest water outlet** in the system. (G15.23) This is in addition to the double check valve arrangement inside the building.

Mini irrigation systems and porous hoses used in a house garden require a double check valve as minimum protection, in addition to **a pipe interrupter with moving element (Type DB) fitted at least 300mm above the highest water outlet** in the system. (G15.23) This is illustrated below.

Protection from hosepipe used with mini irrigation systems and porous hoses where the ground is level or slopes away from the building. (House garden only)

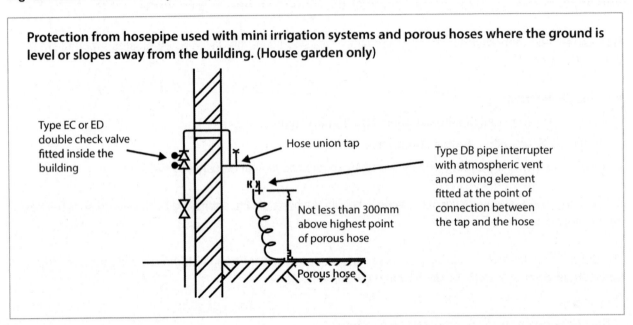

Type EC or ED double check valve fitted inside the building

Hose union tap

Type DB pipe interrupter with atmospheric vent and moving element fitted at the point of connection between the tap and the hose

Not less than 300mm above highest point of porous hose

Porous hose

Protection from hosepipe used with mini irrigation systems and porous hoses where the ground rises away from the building. (House garden only)

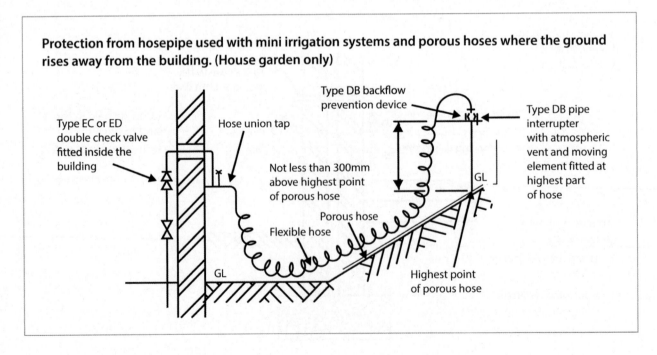

Type EC or ED double check valve fitted inside the building

Hose union tap

Type DB backflow prevention device

Type DB pipe interrupter with atmospheric vent and moving element fitted at highest part of hose

Not less than 300mm above highest point of porous hose

GL

Porous hose

Flexible hose

GL

Highest point of porous hose

Are existing garden taps treated differently to those newly installed?

Yes they are! What we have said up to now has been aimed mainly at new garden tap installations. We'll look at existing taps now!

Existing garden taps in house gardens

While the Water Byelaws do not apply retrospectively, appropriate steps must be taken against any known situation where there is a potential risk of backflow from hoses. In theory, if the tap was fitted legally under the previous Byelaws **and still has a hose connected to it**, it remains legal. However, as soon as the hose is disconnected and reconnected then the installation becomes illegal unless appropriate steps are taken.

Where an outside tap is being replaced (or a hose is being reconnected to a previous installation) then the following steps apply:

(a) if practicable, **a double check valve Type EC or ED** should be provided on the supply to the tap installed inside the building as shown on page 47; or

(b) where it is not practicable to locate the double check valve within the building, the **tap** could be replaced with one with either a **hose union tap that incorporates a double check valve (Type HUK1)** or **a hose union backflow preventer (Type HA) or a double check valve (Type EC or ED) fitted and permanently secured to the outlet of the tap.**

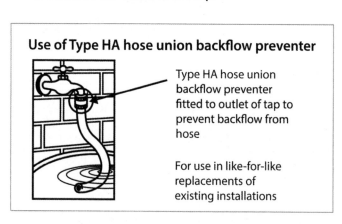

Use of Type HA hose union backflow preventer

Type HA hose union backflow preventer fitted to outlet of tap to prevent backflow from hose

For use in like-for-like replacements of existing installations

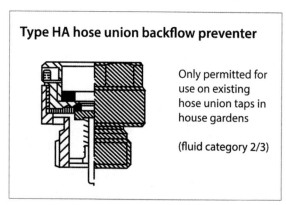

Type HA hose union backflow preventer

Only permitted for use on existing hose union taps in house gardens

(fluid category 2/3)

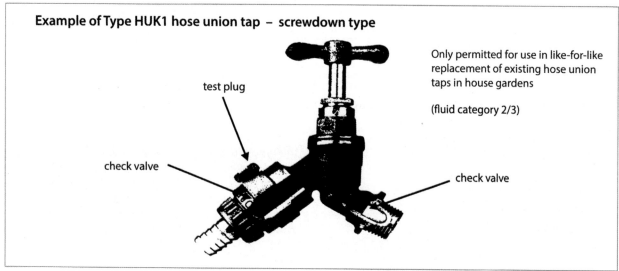

Example of Type HUK1 hose union tap – screwdown type

test plug

check valve

check valve

Only permitted for use in like-for-like replacement of existing hose union taps in house gardens

(fluid category 2/3)

What about outside taps and systems in commercial premises?

Taps used for non-domestic applications generally **present a higher risk than** those **in domestic premises**, backflow protection should be provided to suit the level of risk and the application and where appropriate, a zone protection system (G15.18) e.g. Commercial, horticultural or industrial applications.

For example **soil watering and irrigation systems such as permeable hoses, where the water discharge point/s are less than 150mm above the soil,** are considered to be a **fluid category 5 risk** and should be only supplied through one of the following backflow prevention devices: (G15.19)

– *Type AA air gap with unrestricted discharge*
– *Type AB Air gap with weir overflow*
– *Type AD Air gap with injector*
– *Type AUK 1 air gap arrangement*

Backflow protection in agricultural and horticultural installations

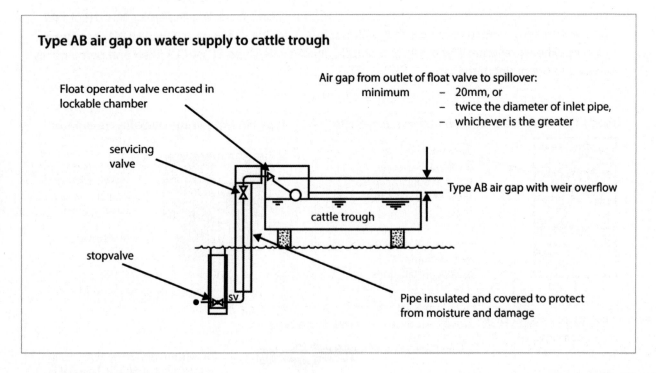

Type AB air gap on water supply to cattle trough

Float operated valve encased in lockable chamber

Air gap from outlet of float valve to spillover:
minimum   – 20mm, or
– twice the diameter of inlet pipe,
– whichever is the greater

servicing valve

Type AB air gap with weir overflow

cattle trough

stopvalve

SV

Pipe insulated and covered to protect from moisture and damage

Animal drinking bowls with air gap at appliance

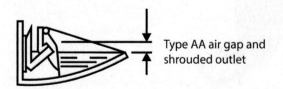

Type AA air gap and shrouded outlet

This type with spring return valve or float valve operated when depressed by animals mouth **may be connected directly from a supply pipe or distributing pipe providing the Type AA air gap is maintained and animals mouth cannot come into contact with the outlet nozzle**

Animal drinking bowls with submerged water outlet or inadequate air gap at appliance

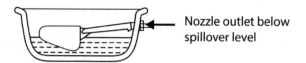

Nozzle outlet below spillover level

Where outlet nozzle is below spillover level or is likely to be contaminated by animals mouth the bowl must be supplied from a dedicated distributing pipe that only supplies similar appliances

Example of interposed cistern used to protect against contamination from cattle drinking bowls

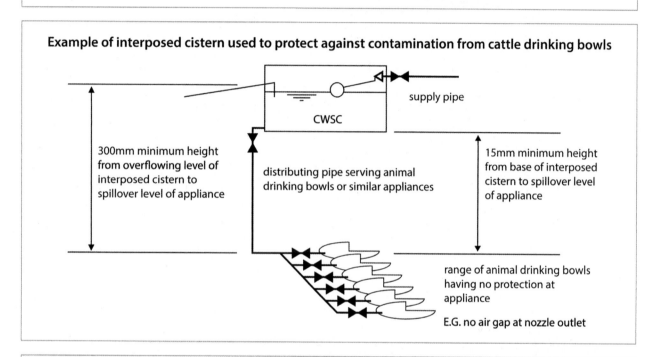

supply pipe

CWSC

300mm minimum height from overflowing level of interposed cistern to spillover level of appliance

distributing pipe serving animal drinking bowls or similar appliances

15mm minimum height from base of interposed cistern to spillover level of appliance

range of animal drinking bowls having no protection at appliance

E.G. no air gap at nozzle outlet

Filling a vessel used to store or mix chemical fertilisers

a) **fill method using Type AA air gap for fluid category 5 contamination risk**

b) **fill method using Type DB pipe interrupter with vent and moving element for fluid category 4 contamination risk**

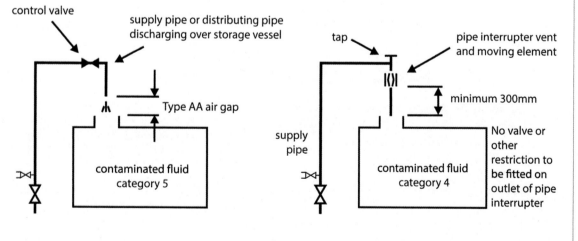

control valve

supply pipe or distributing pipe discharging over storage vessel

Type AA air gap

contaminated fluid category 5

tap

pipe interrupter vent and moving element

minimum 300mm

supply pipe

contaminated fluid category 4

No valve or other restriction to be fitted on outlet of pipe interrupter

Backflow protection in industrial and commercial installations

Type AA air gap to protect supply pipe delivering water to a cistern or vessel used to store or mix contaminated liquid

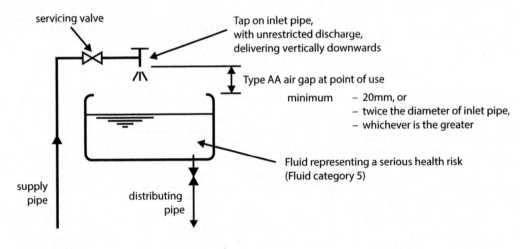

servicing valve

Tap on inlet pipe, with unrestricted discharge, delivering vertically downwards

Type AA air gap at point of use

minimum   – 20mm, or
          – twice the diameter of inlet pipe,
          – whichever is the greater

Fluid representing a serious health risk
(Fluid category 5)

supply pipe

distributing pipe

Type AB air gap to feed an expansion cistern in commercial or industrial premises

Also suitable for use with quality process water e.g. dental surgeries

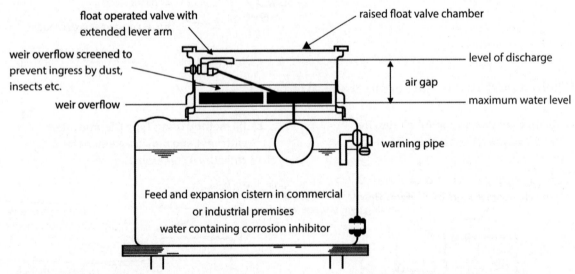

float operated valve with extended lever arm

raised float valve chamber

weir overflow screened to prevent ingress by dust, insects etc.

level of discharge

air gap

weir overflow

maximum water level

warning pipe

Feed and expansion cistern in commercial or industrial premises
water containing corrosion inhibitor

Air gap from outlet of float valve to critical water level:
minimum – 20mm, or
       – twice the diameter of inlet pipe,
       – whichever is the greater

Type AB air gap is the vertical distance between the lowest point of discharge into the vessel and the maximum water level expected during fault conditions e.g. float valve jammed in fully open position

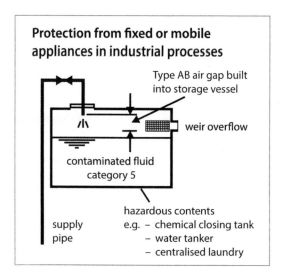

Protection from fixed or mobile appliances in industrial processes

Type AB air gap built into storage vessel

weir overflow

contaminated fluid category 5

supply pipe

hazardous contents
e.g. – chemical closing tank
– water tanker
– centralised laundry

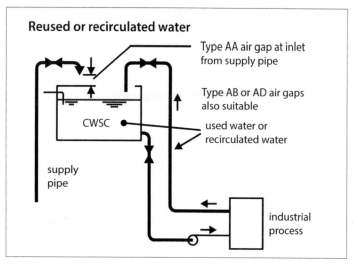

Reused or recirculated water

Type AA air gap at inlet from supply pipe

Type AB or AD air gaps also suitable

CWSC

used water or recirculated water

supply pipe

industrial process

What is whole site and zone protection?

It is what we used to know as 'secondary protection' and is used to protect one building from another or one part of a building from another part.

G15.24 says that **whole site or zone protection should be provided where:**

(a) **a supply pipe or distributing pipe conveys water to two or more separately occupied premises**; or

(b) **a supply pipe conveys water to premises that are required to provide sufficient water storage for 24 hours ordinary use**.

The protection device used **will depend on the level of risk** as judged by the water undertaker.

Whole site or zone protection should be provided in addition to any point of use protection devices required within the system (G15.25).

Zone protection is particularly important in premises where industrial, medical or chemical processes are undertaken. (G15.26)

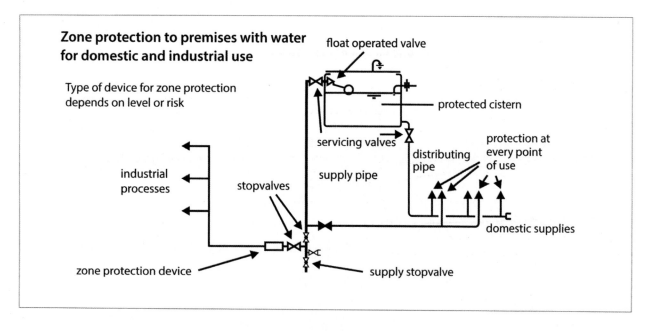

Zone protection to premises with water for domestic and industrial use

float operated valve

Type of device for zone protection depends on level or risk

protected cistern

servicing valves

protection at every point of use

industrial processes

distributing pipe

stopvalves

supply pipe

domestic supplies

zone protection device

supply stopvalve

Does backflow protection have to be applied to fire systems?

Yes! Backflow protection should be applied to suit the level of risk.

Wet sprinkler systems (without additives), first aid fire hose reels, and hydrant landing valves are fluid category 2 risk and require the minimum protection of **a single check valve only**. (G15.27)

Wet sprinkler systems with additives to prevent freezing, and systems containing hydro pneumatic pressure vessels are considered to be in fluid risk category 4 (G15.27/28), and will require either a verifiable backflow preventer (RPZ valve) or be fitted with a suitable air gap such as Type AA, AB, AD or AUK1.

Where a common supply pipe serves a fire protection system, and a supply pipe for drinking water and domestic purposes, the fire supply should be connected immediately on entry to the building and appropriate backflow protection should be fitted close to the point of connection. (15.29)

Water stored for fire protection purposes is considered to be a fluid level 5 risk (Table 6.1e) except that an elevated, covered and protected cistern **may** be permitted to serve both water for domestic purposes and fire sprinkler outlets, provided no antifreeze additives are used. (See diagram *R15.27B in the* Water Regulations Guide)

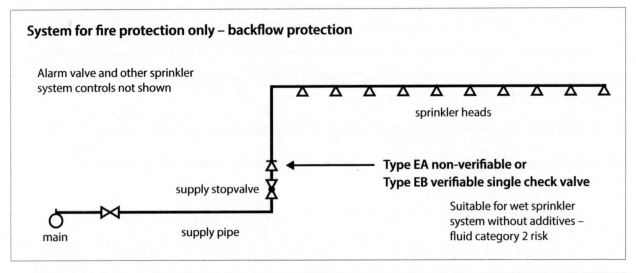

System for fire protection only – backflow protection

Alarm valve and other sprinkler system controls not shown

sprinkler heads

Type EA non-verifiable or Type EB verifiable single check valve

supply stopvalve

Suitable for wet sprinkler system without additives – fluid category 2 risk

main   supply pipe

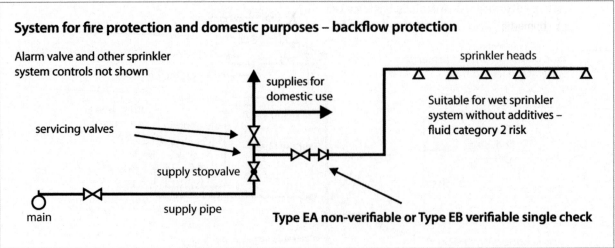

System for fire protection and domestic purposes – backflow protection

Alarm valve and other sprinkler system controls not shown

sprinkler heads

supplies for domestic use

Suitable for wet sprinkler system without additives – fluid category 2 risk

servicing valves

supply stopvalve

main   supply pipe

Type EA non-verifiable or Type EB verifiable single check

Stored water for fire protection purposes – backflow protection

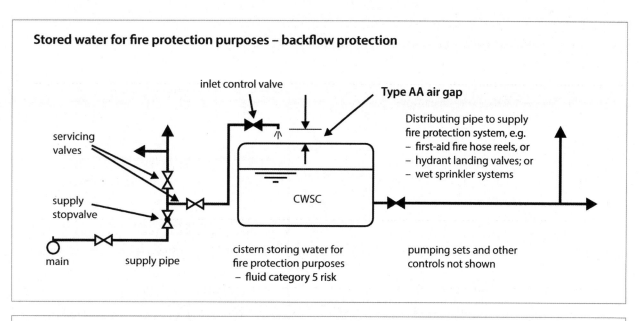

inlet control valve

Type AA air gap

servicing valves

supply stopvalve

CWSC

Distributing pipe to supply fire protection system, e.g.
– first-aid fire hose reels, or
– hydrant landing valves; or
– wet sprinkler systems

main

supply pipe

cistern storing water for fire protection purposes
– fluid category 5 risk

pumping sets and other controls not shown

Combined sprinkler and water supply system for dwellings and similar premises – backflow protection

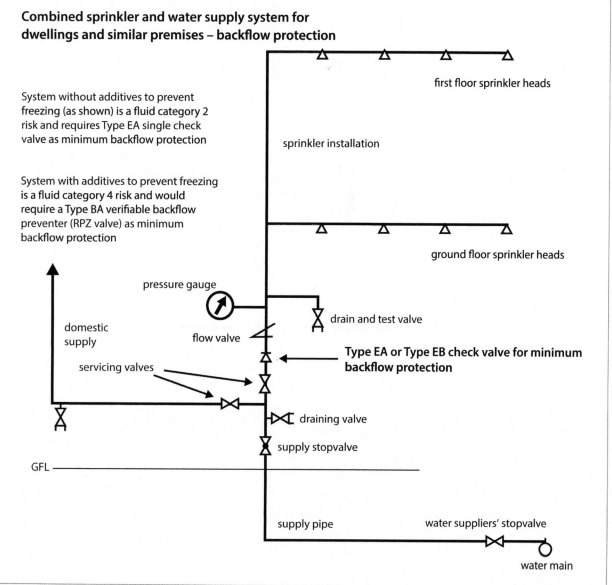

first floor sprinkler heads

System without additives to prevent freezing (as shown) is a fluid category 2 risk and requires Type EA single check valve as minimum backflow protection

sprinkler installation

System with additives to prevent freezing is a fluid category 4 risk and would require a Type BA verifiable backflow preventer (RPZ valve) as minimum backflow protection

ground floor sprinkler heads

pressure gauge

drain and test valve

domestic supply

flow valve

Type EA or Type EB check valve for minimum backflow protection

servicing valves

draining valve

supply stopvalve

GFL

supply pipe

water suppliers' stopvalve

water main

**Boosted system for fire protection using hydro-pneumatic pressure vessel
– backflow protection**

Alarm valve and other sprinkler system controls not shown

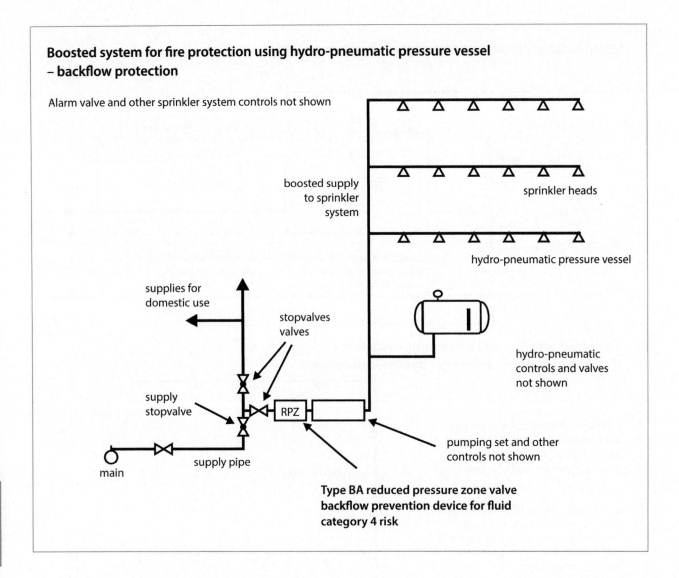

boosted supply
to sprinkler
system

sprinkler heads

hydro-pneumatic pressure vessel

supplies for
domestic use

stopvalves
valves

supply
stopvalve

RPZ

hydro-pneumatic
controls and valves
not shown

pumping set and other
controls not shown

supply pipe

main

**Type BA reduced pressure zone valve
backflow prevention device for fluid
category 4 risk**

Self-assessment questions (4)

1. Complete the following sentences relating to outside taps

    a) Hosepipes held in the hand for garden and other uses should be fitted with a _____

    _____ _____ at the hose outlet.

    b) In a house situation, any garden tap to which a hose can be connected should be fitted with a

    _____ _____ _____

    positioned _____ the building where it will not be subjected to frost damage.

2. State the recommended backflow protection required for a hose union tap in the following situations where the hose is to be hand held and is to be used on domestic premises for the spraying of fertilisers and household detergents.

    _____

3. The following diagram illustrates a garden tap in domestic premises.

    a) Complete the annotations to show the recommended backflow protection.

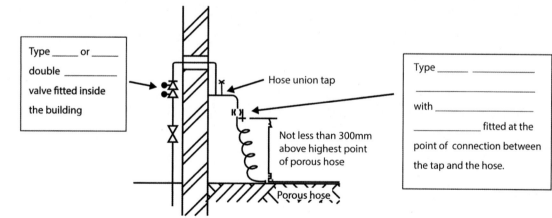

Protection from hosepipe used with mini irrigation systems and porous hoses where the ground is level or slopes away from the building. (House garden only)

Type _____ or _____
double _____
valve fitted inside
the building

Hose union tap

Not less than 300mm
above highest point
of porous hose

Porous hose

Type _____ _____
_____
with _____
_____ fitted at the
point of connection between
the tap and the hose.

    b) Briefly describe how the above installation of a porous hose would be different if the ground was rising away from the tap connection.

    _____

    _____

    _____

4. Where existing outside taps fitted under previous Byelaws are being replaced, then they have their backflow prevention decide fitted outside the building.

a) _____

_____

b) _____

_____

c) _____

_____

5. Taps in commercial, industrial and horticultural premises present a more serious backflow risk than those in domestic premises. State the level of risk for a hosepipe installation in an industrial situation, and give ONE permitted backflow prevention device or arrangement to protect against that risk.

The level of risk is _____

The device or arrangement could be _____

_____

6. State ONE suitable backflow device or arrangement for each of the following:

a) a cattle drinking trough

_____

b) an animal drinking bowl supplied from a supply pipe

_____

c) an animal drinking bowl having an outlet which is below the spillover level of the bowl

_____

7. State ONE suitable backflow protection device or arrangement to protect against the following two applications in industrial or agricultural situations and complete the annotations in the diagrams below.

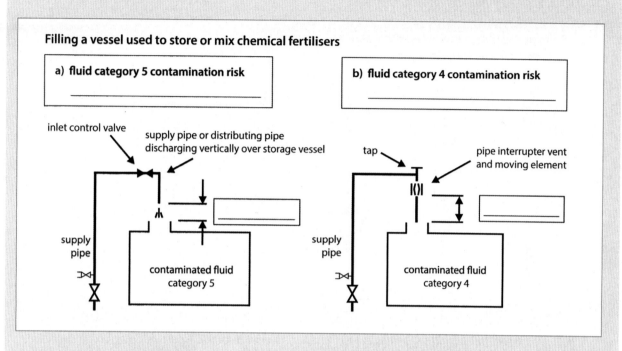

Filling a vessel used to store or mix chemical fertilisers

a) **fluid category 5 contamination risk**
_____

b) **fluid category 4 contamination risk**
_____

inlet control valve

supply pipe or distributing pipe discharging vertically over storage vessel

supply pipe

contaminated fluid category 5

tap

pipe interrupter vent and moving element

supply pipe

contaminated fluid category 4

8. Complete the following sentences relating to whole site and zone protection.

a) Whole site or zone protection should be provided where:

i)   a supply pipe or distributing pipe conveys water to _____ or

_____ separately occupied premises; or

ii) a supply pipe conveys water to premises that are required to provide sufficient water storage for

_____   _____ ordinary use.

b) The protection device used will depend on the level of risk as judged by the

_____   _____

c) Whole site or zone protection should be provided in _____ to any point of use protection devices required within the system

9. Indicate with an arrow in the diagram below, the correct position for a zone protection device to be fitted.

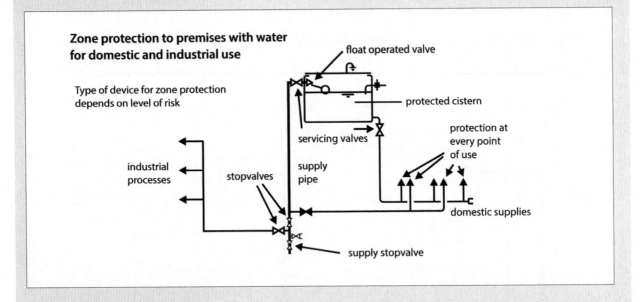

10. Fire fighting systems are required to have backflow prevention devices fitted according to the risk. On the following three diagrams indicate in the spaces provided, the fluid risk category, a suitable backflow prevention device for the installation, and show with an arrow where the backflow device should be positioned.

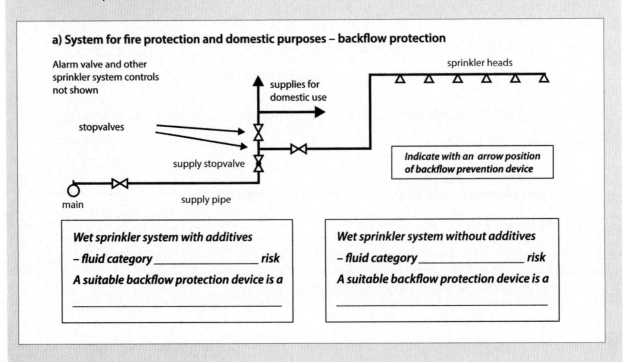

b) Stored water for fire protection purposes – backflow protection

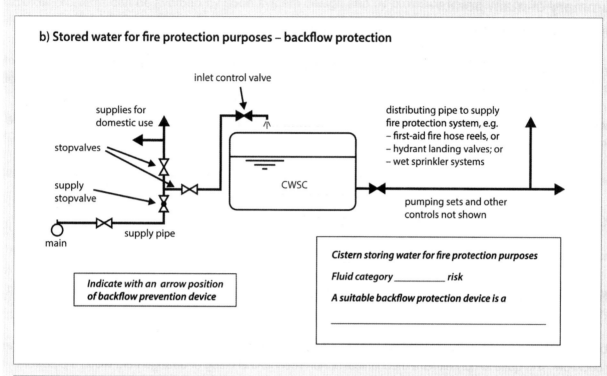

Indicate with an arrow position of backflow prevention device

Cistern storing water for fire protection purposes

Fluid category _____ risk

A suitable backflow protection device is a

_____

c) Boosted system for fire protection using hydro-pneumatic pressure vessel – backflow protection

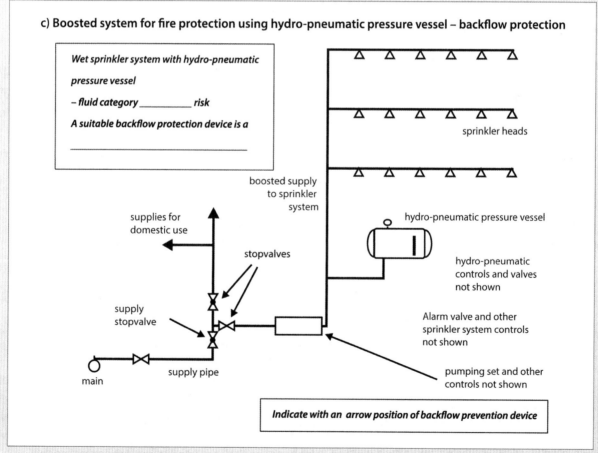

Wet sprinkler system with hydro-pneumatic pressure vessel

– fluid category _____ risk

A suitable backflow protection device is a

_____

Indicate with an arrow position of backflow prevention device

Check your answers on pages 74 –78.

Both **hot and cold connections** should be taken **from independent** dedicated **distributing pipes that do not supply other appliances;** except when:

(a) the common distributing pipe **serves only the bidet and a WC or urinal flushing cistern**

(b) the bidet is the **lowest fitting served** and there is no likelihood of other fittings being connected at a lower level at a later date, and the **connection to the common distributing pipe is not less than 300mm above the spillover level of the bowl,** or

(c) in the case of an over-rim bidet with a flexible spray connection, not less than 300mm above the spillover level of any appliance that the spray outlet might reach.

Single outlet taps, combination taps and fixed shower heads should discharge above the appliance and terminate **with a Type AUK 2 tap gap.**

Sinks in both domestic and non-domestic situations **are a fluid category 5 backflow risk** and the minimum protection is the Type AUK3 air gap.

Where appliances such as baths and wash basins in domestic premises have submerged inlets they are considered to be a category 3 risk and both **hot and cold inlets should be supplied through Type EC, or ED double check valves.**

Household washing machines and dishwashers should have backflow protection to **fluid level 3** built in during manufacture. They will be listed in the Water Materials and Fittings Directory if approved under the Water Regulations Advisory Scheme.

Where a connection hose is not approved, a check valve is required to prevent backflow from the hose.

Commercial washing machines in laundromats or similar premises are a category 4 risk, whilst clothes washing plant or equipment **in health care establishments are fluid category 5.**

Drinking water fountains should have a minimum 25mm air gap between the water delivery nozzle and the spillover level of the bowl. Additionally **the nozzle should be screened or shrouded to prevent mouth contact.**

Outside taps and garden supplies.

Hosepipes held in the hand for garden and other uses **should be fitted with a self-closing mechanism at the hose outlet.** (G15.17)

In a house situation:
– **any garden tap to which a hose connection can be made**
– **should be fitted with a double checkvalve**
– **positioned inside a building** where it will not be subjected to frost damage (G15.20).

A double check valve is considered **sufficient protection for hand held hoses used for spraying fertilisers or domestic detergents in house garden situations.**

For the spraying of insecticides (fluid category 5) suitable protection would be **a pipe interrupter with moving element (Type DB) fitted at the connection of the hose and at least 300mm above the highest water outlet** in the system. (G15.23) This is in addition to the double check valve arrangement inside the building.

Mini irrigation systems and porous hoses used in a house garden require a double check valve as minimum protection, in addition to **a pipe interrupter with moving element (Type DB) fitted at least 300mm above the highest water outlet** in the system. (G15.23)

Existing garden taps in house gardens

While the Water Byelaws do not apply retrospectively, appropriate steps must be taken against any known situation where there is a potential risk of backflow from hoses. In theory, if the tap was fitted legally under the previous Byelaws **and still has a hose connected to it**, it remains legal. However, as soon as the hose is disconnected and reconnected then the installation becomes illegal unless appropriate steps are taken.

Where an outside tap is being replaced (or a hose is being reconnected to a previous installation) then the following steps apply:

(a) if practicable, **a double check valve Type EC or ED** should be provided on the supply to the tap installed inside the building as shown on page 45; or

(b) where it is not practicable to locate the double check valve within the building, the **tap** could be replaced with one with either a **hose union tap that incorporates a double check valve (Type HUK1)** or a **hose union backflow preventer (Type HA) or a double check valve (Type EC or ED) fitted and permanently secured to the outlet of the tap.**

Taps for non-domestic applications generally present a higher risk than those in domestic premises and should be provided with backflow protection to suit the level of risk and the application (G15.18). e.g. commercial, horticultural or industrial applications. Additionally, **there may be a need to also provide zone protection.** (G15.18)

For example **soil watering and irrigation systems such as permeable hoses, where the water discharge point/s are less than 150mm above the soil, are considered to be a fluid category 5 risk.** (G15.19)

Whole site and zone protection, previously referred to as 'secondary backflow protection' is used to protect one building from another or one part of a building from another part.

G15.24 says that **whole site or zone protection should be provided where**:

(a) **a supply pipe or distributing pipe conveys water to two or more separately occupied premises;** or

(b) **a supply pipe conveys water to premises that are required to provide sufficient water storage for 24 hours ordinary use.**

The protection device used **will depend on the level of risk** as judged by the water undertaker.

Whole site or zone protection should be provided in addition to any point of use protection devices required within the system. (G15.25)

Zone protection is particularly important in premises where industrial, medical or chemical processes are undertaken. (G15.26)

Fire protection systems

Wet sprinkler systems (without additives), first aid fire hose reels, and hydrant landing valves are fluid category 2 risk and require the minimum protection of a single check valve. (G15.27)

Wet sprinkler systems with additives to prevent freezing, and systems incorporating hydro-pneumatic pressure vessels are in fluid risk category 4 (G15.27/28), and will require either a Type BA verifiable backflow preventer (RPZ valve) or be fitted with a suitable air gap such as Type AA, AB, AD or AUK1.

Where a common supply pipe serves a fire protection system, and a supply pipe for drinking water and domestic purposes, the fire supply should be connected immediately on entry to the building and appropriate backflow protection should be fitted close to the point of connection. (G15.29)

Water stored for fire protection purposes is considered to be a fluid level 5 risk. (Table 6.1e)

Answers to self-assessment questions (1)

1. Give a simple description of backflow.

   *Backflow can be described as '...**the flow of water in a direction contrary to the intended normal direction of flow...**',*

2. The following two diagrams illustrate backpressure and backsiphonage. Complete the titles to show which type of backflow is represented.

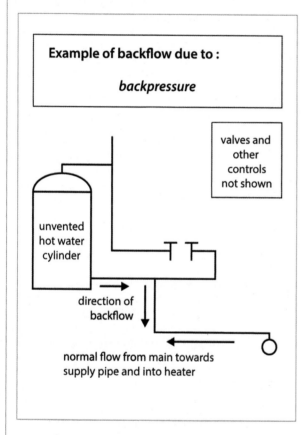

Example of backflow due to :

backpressure

valves and other controls not shown

unvented hot water cylinder

direction of backflow

normal flow from main towards supply pipe and into heater

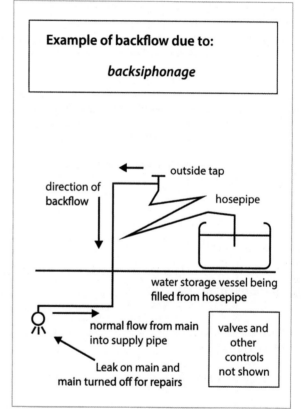

Example of backflow due to:

backsiphonage

direction of backflow

outside tap

hosepipe

water storage vessel being filled from hosepipe

normal flow from main into supply pipe

valves and other controls not shown

Leak on main and main turned off for repairs

3. Describe backpressure and backsiphonage.

   *Backpressure can be described as '**the reversal of flow in a pipe caused by an increase in pressure in the system**'.*

   *Backsiphonage is; '**backflow caused by siphonage of water from a cistern or appliance back into the pipe which feeds it**'.*

4. Schedule 1 of the Water Supply (Water Fittings) Regulations 1999 sets out five fluid categories which are shown in the following chart. Complete the chart to show the severity of risk in each case.

Water Regulations – fluid risk categories

5	*Serious* health hazard
4	*Significant* health hazard
3	*Slight* health hazard
2	Aesthetic quality is impaired
1	No health hazard or impairment in its quality

5. Most backflow devices are 'point of use' devices. What is meant by a 'point of use' device?

A term 'point of use' backflow device is one that is **positioned at or near to the point where the water is supplied for use.**

6. Complete the following statements giving general advice on the installation of backflow prevention devices in G15.7.

Backflow prevention devices:

a) *should be* **readily accessible** *for inspection, operational maintenance and renewal*

b) *for risk categories 2 and 3, except Types HA and HUK1, devices should* **NOT** *be located outside the building*

c) *should not be buried in the* **ground**

d) *of the vented or verifiable type should NOT be installed in chambers below* **ground level** *or where liable to* **flooding**

e) *used for category 4 devices, should have* **line strainers** *fitted upstream (before the backflow device) and a* **servicing valve** *upstream of the strainer*

f) *of the reduced pressure zone valve type, should terminate with a Type* **AA** *air gap located at least* **300mm** *above ground or floor level*

Answers to self-assessment questions (2)

1. The following table gives a list of non-mechanically operated backflow prevention devices. Using Table S6.1 on page 8, complete the table to show the minimum category of risk for which each backflow risk is suited.

Type of device	Description of arrangement or device	Minimum category of backpressure risk to which it is suited	Minimum category of backsiphonage risk to which it is suited
Type AA	Air gap with unrestricted discharge (formerly Type A air gap)	5	5
Type AB	Air gap with weir overflow	5	5
Type AG	Air gap with minimum size circular overflow (satisfies requirements of Type B air gap to BS 6281 Part 2)	3	3
Type AUK1	Air gap with interposed cistern	3	5
Type DC	Pipe interrupter with permanent atmospheric vent	not applicable	5

2. The following table gives a list of mechanically operated backflow prevention devices. Using Table S6.2 on page 9 complete the table to show the minimum category of risk for which each backflow device is suited.

Type of device	Description of arrangement or device	Minimum category of backpressure risk to which it is suited	Minimum category of backsiphonage risk to which it is suited
Type BA	Verifiable backflow preventer with reduced pressure zones	4	4
Type DB	Pipe interrupter with atmospheric vent and moving element	not applicable	4
Type EA	Verifiable single check valve	2	2
Type EB	Non-verifiable single check valve	2	2
Type EC	Verifiable double check valve	3	3
Type ED	Non-verifiable double check valve	3	3
Type HUK1	Hose union tap which incorporates a double check valve	3	3

3. A Type AA air gap should have an unrestricted discharge delivering vertically downwards and its dimension depends on the size of the inlet to the vessel served.

   Complete the following rules relating to a Type AA air gap.

   The Type AA air gap at point of use should be:
   - minimum – **20mm, or**
   - **– twice the diameter of the inlet pipe,**
   - *– whichever is the greater.*

4. Complete the following table of Air gaps at taps, valves and fittings.

Situation	Nominal size of inlet of tap, valve or fitting	Vertical distance between tap or valve outlet and spillover level of receiving appliance (mm)
Domestic situations fluid category 2/3 device AUK2	Up to and including G½	20
	Over G½ and up to G¾	25
	Over G¾	70
Non-domestic situations fluid category 4/5 device AUK3	any size inlet pipe	minimum diameter **20**mm, or twice the diameter of the inlet pipe, whichever is the **greatest**

Note: AUK2 and AUK3 devices are suitable for protection against backsiphonage risk only

Answers to self-assessment questions (3)

1. To make backflow precautions effective in flushing cisterns and devices, there are rules applied to the installation of them. Complete the following diagrams to show correct minimum dimensions for effective backflow prevention.

WC pan supplied from interposed cistern (flushing cistern)

Interposed cistern to include Type AG air gap

> Cistern overflowing level to be at least **300mm** above spillover level of WC pan.
>
> Base of cistern to be at least **15mm** above spillover level of WC pan.

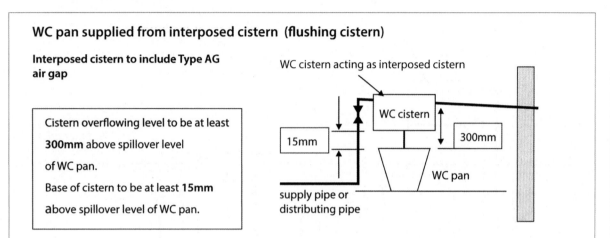

2. Give the correct backflow protection device to suit an over-rim type bidet.

   a) in domestic premises   **Type AUK2 air gap**

   b) in a hospital   **Type AUK3 air gap**

WCs and urinals supplied from pressure flushing valves

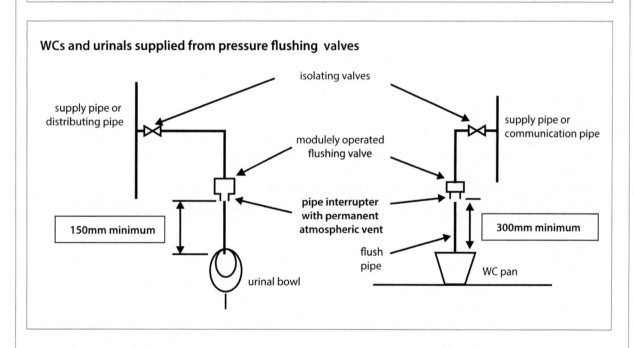

3. Are ascending spray type bidets permitted to be connected directly from the supply pipe?

   Answer: Yes ☐   No ☑

4. Complete the following diagram to show an acceptable method of connecting an ascending spray type bidet. Also complete the labelling to give precautions that should be applied to the installation.

Ascending spray type bidet using dedicated distributing pipes

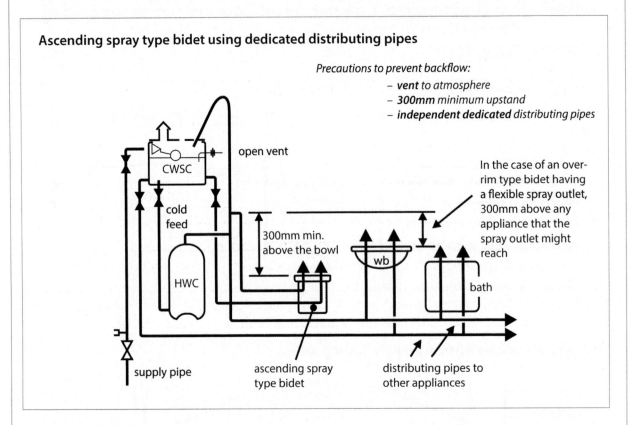

Precautions to prevent backflow:
- *vent* to atmosphere
- *300mm* minimum upstand
- *independent dedicated* distributing pipes

open vent

In the case of an over-rim type bidet having a flexible spray outlet, 300mm above any appliance that the spray outlet might reach

CWSC

cold feed

300mm min. above the bowl

wb

HWC

bath

supply pipe

ascending spray type bidet

distributing pipes to other appliances

5. Taps and showers need different levels of backflow protection depending on their risk assessment. In each of the following cases, state the fluid risk category and the minimum protection needed.

Application of backflow protection to taps and shower fittings	fluid risk category	minimum suitable backflow protection needed
single outlet and combination taps, and shower heads to wash basins and baths in domestic premises	3	Type AUK2 air gap
single outlet and combination taps, to sinks in domestic and commercial premises	5	Type AUK3 air gap
single outlet and combination taps to baths and basins in hospital premises	5	Type AUK3 air gap

6. Baths and basins in domestic premises are a fluid category 3 risk whilst sinks are category 5.

    a) State the usual backflow protection given to:

        i)   *a domestic bath or basin*   **Type AUK2 air gap**

        ii)   *a sink in domestic premises*   **Type AUK3 air gap**

    b) State suitable backflow protection for a basin, bath or shower where the water supply outlet from tap or shower head is below, or could be placed below the spillover level of the appliance.

        A double check valve arrangement on each inlet pipe

7. Washing machines and dishwashers in domestic premises should have protection built in during manufacture. However, if you come across a domestic machine with connection hoses that are not WRAS approved, what protection should be given to the supply or distributing pipe.

    It should have a single check valve immediately before the hosepipe connection

8. Commercial washing machines are a more serious risk than domestic ones. Give a suitable backflow protection device for a range of commercial washing machines in a laundromat.

    AA air gap – BA (RPZ) valve assembly

9. Complete the following diagram to show two important contamination requirements for a drinking water fountain.

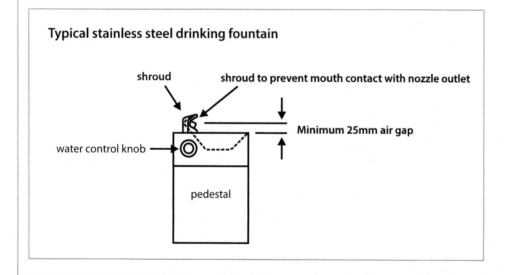

Typical stainless steel drinking fountain

shroud

shroud to prevent mouth contact with nozzle outlet

Minimum 25mm air gap

water control knob

pedestal

Answers to self-assessment questions (4)

1. Complete the following sentences relating to outside taps

   a) Hosepipes held in the hand for garden and other uses should be fitted with a **self closing mechanism** at the hose outlet.

   b) In a house situation any garden tap to which a hose can be connected should be fitted with a **double check valve** positioned **inside** the building where it will not be subjected to frost damage.

2. State the recommended backflow protection required for a hose union tap, where the hose is to be hand held and is to be used on domestic premises for the spraying of fertilisers and household detergents.

   double check valve

3. The following diagram illustrates a garden tap in domestic premises.

   a) Complete the annotations to show the recommended backflow protection.

Protection from hosepipe used with mini irrigation systems and porous hoses where the ground is level or slopes away from the building. (House garden only)

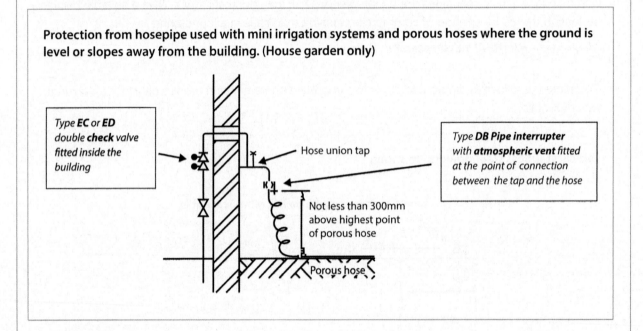

Type **EC** or **ED** double **check** valve fitted inside the building

Hose union tap

Type **DB Pipe interrupter** with **atmospheric vent** fitted at the point of connection between the tap and the hose

Not less than 300mm above highest point of porous hose

Porous hose

---

   b) Briefly describe how the above installation of a porous hose would be different if the ground was rising away from the tap connection.

   The pipe interrupter would be fitted at the end of the hose or above its highest discharge point

4. Where existing outside taps under previous Byelaws are being replaced, then they may have their backflow prevention device fitting outside the building. Give THREE ways in which acceptable protection can be provided.

   a) *a Type EC or ED double check valve installed inside the building*

   b) *a tap that incorporates a Type EC or ED double check valve arrangement*

   c) *a hose union backflow preventer (Type HA), or a Type EC or ED double check valve fitted to the outlet of the tap*

5. Taps in commercial, industrial and horticultural premises present a more serious backflow risk than those in domestic premises. State the level of risk for a hosepipe installation in an industrial situation, and give ONE permitted backflow prevention device or arrangement to protect against that risk.

   The level of risk is **fluid category 5**

   One device or arrangement could be a **Type AUK3 air gap plus zone protection**

   alternatives include: **Type AA air gap**

                              Type AUK1 air gap with interposed cistern

6. State ONE suitable backflow device or arrangement for each of the following:

   a) *a cattle drinking trough*

      Type AB air gap with weir overflow

   b) *an animal drinking bowl supplied from a supply pipe*

      Type AUK3 air gap or Type AA air gap

   c) *an animal drinking bowl having an outlet which is below the spillover level of the bowl*

      Type AUK1 air gap with interposed cistern

7. State ONE suitable backflow protection device or arrangement to protect against the following two applications in industrial or agricultural situations and complete the annotations in the diagrams below.

Filling a vessel used to store or mix chemical fertilisers

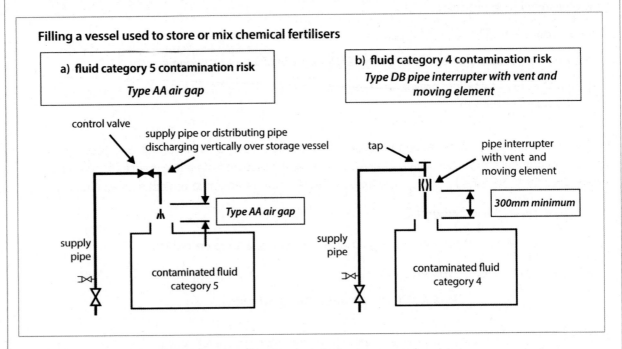

8. Complete the following sentences relating to whole site and zone protection.

a) *Whole site or zone protection should be provided where:*

   i) *a supply pipe or distributing pipe conveys water to* **two** *or* **more** *separately occupied premises; or*

b) ii) *a supply pipe conveys water to premises that are required to provide sufficient water storage for* **24 hours** *ordinary use.*

c) *The protection device used will depend on the level of risk as judged by the* **water undertaker**

d) *Whole site or zone protection should be provided in* **addition** *to any point of use protection devices required within the system*

9. Indicate with an arrow in the diagram below, the correct position for a zone protection device to be fitted.

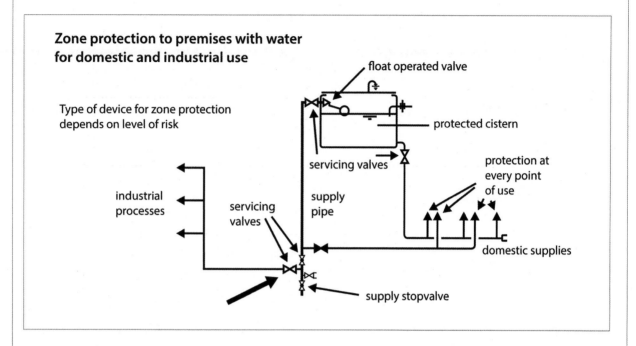

10. Fire fighting systems are required to have backflow prevention devices fitted according to the risk. On the following three diagrams indicate in the spaces provided, the fluid risk category, a suitable backflow prevention device for the installation, and show with an arrow where the backflow device should be positioned.

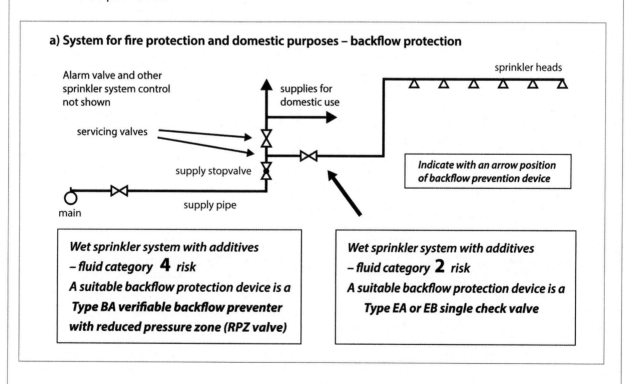

b) Stored water for fire protection purposes – backflow protection

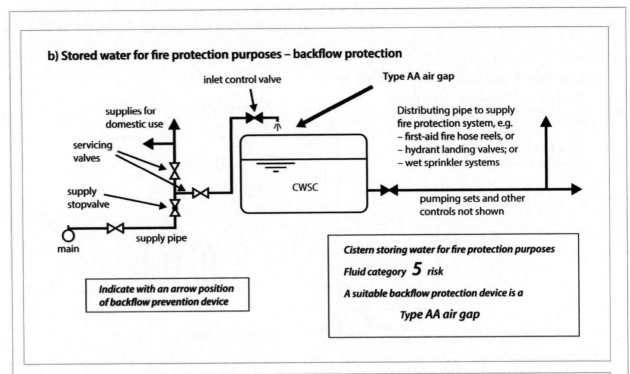

Indicate with an arrow position of backflow prevention device

Cistern storing water for fire protection purposes

Fluid category **5** risk

A suitable backflow protection device is a

Type AA air gap

c) Boosted system for fire protection using hydro-pneumatic pressure vessel – backflow protection

Wet sprinkler system with hydro-pneumatic pressure vessel

– fluid category **4** risk

A suitable backflow protection device is a

RPZ valve

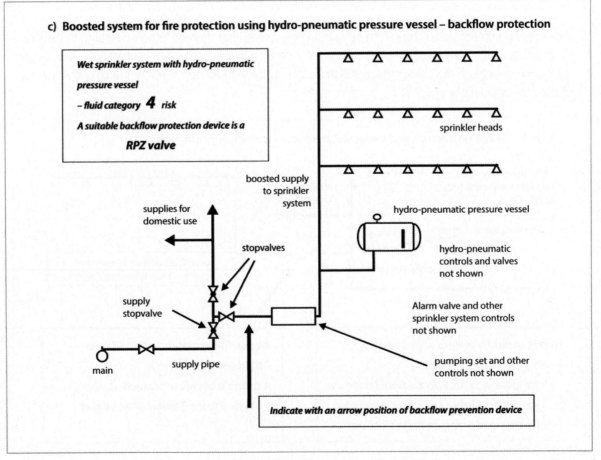

Indicate with an arrow position of backflow prevention device

What to do next

Well done!

That was a marathon wasn't it?

That is the biggest and the most important module finished.

Now it is time to take a look at Cold water services

Please go on to **Module 9**

Water Industry Act 1991:

Water Supply (Water Fittings) Regulations 1999

An Open Learning Course

Module 9

Cold water services

Introduction

This module looks at Schedule 2: Paragraph 16 Cold water services. It's only one paragraph but it covers quite a lot of ground.

Whilst the title is 'cold water services', Paragraph 16 is concerned mainly with cold water storage cisterns including control of incoming water, overflow pipes and warning pipes, and preventing waste and contamination in cisterns.

Here, as in many of the paragraphs of Schedule 2, the over-riding concern is that water supplied by the undertaker for domestic purposes (including drinking water), should remain wholesome.

It looks at the provision of servicing valves on inlet and outlet pipes to cisterns and at the provision of insulation for the prevention of heat losses and heat gains in cisterns.

So! These requirements are aimed at primarily at **preventing contamination of water** in any storage vessel, its distributing pipes and particularly its supply pipe.

Of secondary concern, but still very important, is that cold water storage cisterns should be installed so as to **prevent waste** or undue use of water.

So! Lets get down to business!

What are the Requirements?

Schedule 2: Paragraph 16: Cold water services

16.(1) Every pipe supplying water connected to a storage cistern shall be fitted with an effective adjustable valve capable of shutting-off the inflow of water at a suitable level below the overflowing level of the cistern.

(2) Every inlet to a storage cistern, combined feed and expansion cistern, WC flushing cistern or urinal flushing cistern shall be fitted with a servicing valve on the inlet pipe adjacent to the cistern.

(3) Every storage cistern, except one supplying water to the primary circuit of a heating system shall be fitted with a servicing valve on the outlet pipe.

(4) Every storage cistern shall be fitted with:

(a) an overflow pipe, with a suitable means of warning of an impending overflow, which excludes insects;

(b) a cover positioned so as to exclude light and insects.

(c) thermal insulation to minimise freezing or undue warming.

(5) Every storage cistern shall be so installed as to minimise the risk of contamination of stored water. The cistern shall be of an appropriate size, and the pipe connections to the cistern shall be so positioned, as to allow free circulation and to prevent areas of stagnant water from developing.

What is important about cold water storage cisterns?

There are a number of points to be considered here, inlet and outlet pipes, control valves, warning pipes and overflow pipes, water levels, cistern support, prevention of contamination, materials for cisterns insulation and siting of cisterns.

So let's take it step by step!

First a general look at cisterns and what is required of them.

For cold water services storage cisterns to comply with the requirements of the Water Supply (Water Fittings) Regulations & Scottish Water Byelaws the complete installation must be designed and installed so that:

- it is of an appropriate quality and standard
- the risk of contamination is minimised
- it incorporates essential design features (inlet and outlet arrangements, overflow and warning pipe provision) and correctly installed

The following diagram illustrates a number of points that apply to all cisterns used to supply water for drinking and domestic purposes. Take note of these.

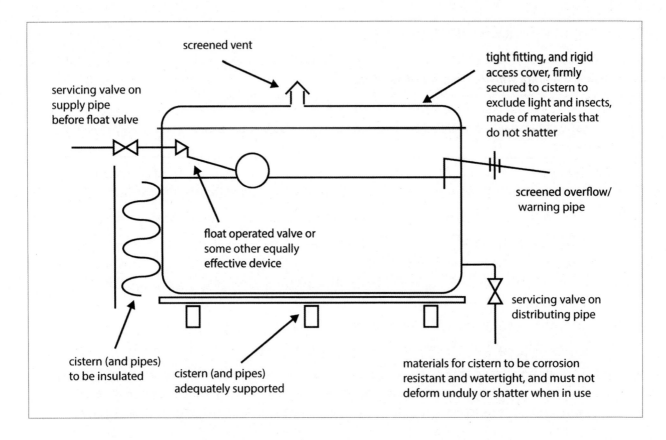

All components making up a cold water storage cistern, including the inlets, outlets and servicing valves must conform to the requirements of Regulation 4.

In general cisterns should be:

(a) fitted with an effective inlet control device to maintain the correct water level;

(b) fitted with servicing valves on inlet and outlet pipes;

(c) fitted with a screened warning/overflow pipe to warn against impending overflow;

(d) covered so to exclude light or insects, and

(e) insulated to prevent heat losses and undue warming;

(f) installed so that risk of contamination is minimised, and

(g) arranged so that water can circulate and stagnation will not occur;

(h) supported to avoid distortion or damage that might cause them to leak;

(i) readily accessible for inspection and cleansing;

(j) all non-metallic materials in contact with the contents, including any surface where condensation forms, must comply with the requirements of BS 6920.

Of the points mentioned most of them we have already dealt with to some degree in previous modules. As you can see there is some overlap in what is covered by some paragraphs of Schedule 2 and in order to understand what is meant by each individual paragraph we also need perhaps to go over some of the ground again.

Cisterns used for domestic purposes

Most cisterns, and particularly those in dwellings, are used in some way to store water for domestic purposes. This means that in addition to the points listed in the diagram on page 4, cisterns for domestic purposes must be arranged so as to prevent contamination from occurring in the cistern water.

You will be familiar with the old 'byelaw 30' cistern requirements, and similar requirements are still valid under the Regulations.

The following diagram illustrates the requirements for cisterns for domestic purposes and cisterns that are used to store drinking water.

These are often termed 'protected' cisterns and contain water of drinking quality conforming to fluid category 1.

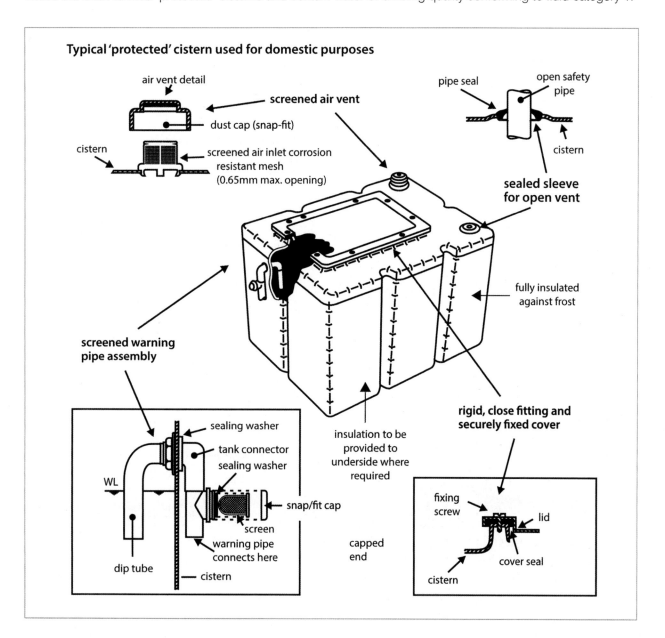

Typical 'protected' cistern used for domestic purposes

Are there any rules relating to cistern connections and control valves?

The simple answer is **Yes!** But you will need a more detailed answer which looks at inlet and outlet control valves, overflow and warning pipes, contamination and frost protection. More details follow.

Cistern inlet controls

Paragraph 16(1) requires **every pipe supplying water to a storage cistern to be fitted with an effective adjustable shut off device** that will close when the water reaches its normal full level, and you must be able to adjust the water level in the cistern.

Normally this device will be a float operated valve such as those illustrated below, but on larger cisterns it could be a float switch control connected to an electrically operated valve or pump.

Where a float operated valve is used in a water storage cistern it should comply with one of the following standards:
–   BS 1212: Part 1 Portsmouth type
–   BS 1212: Part 2 Diaphragm valve made of brass
–   BS 1212: Part 3 Diaphragm valve made of plastic

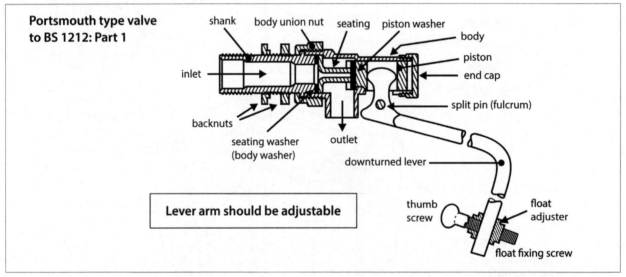

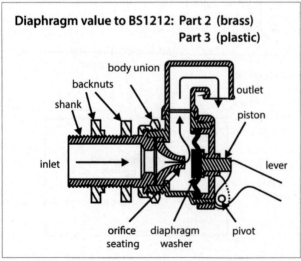

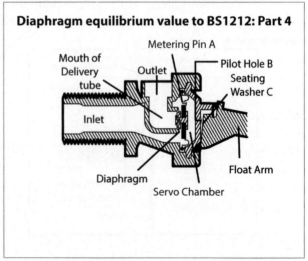

Note: Float operated valves used in WC cisterns should comply with either BS 1212: Parts 2, 3 or 4. The valve to BS 1212: Part 4 is designed for use in WC cisterns.

What about cistern inlet control valves that do not comply with the standards listed above?

There are a number of these! For instance, BS 1212 only covers float valves up to 50mm (2") in diameter.

Another example is the innovative valve shown here which has a ceramic disc valve arrangement and which is WRAS approved.

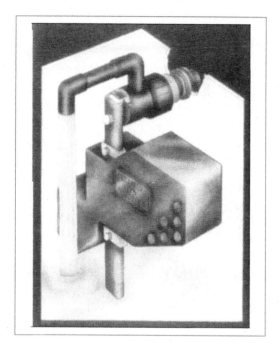

So! If you intend to fit a valve to a cistern that is not to BS 1212 or is of a size larger than 50mm, you will need to make certain that it meets with approval under Water Regulations. You have three options open to you, either:

i)   to look in the Water Fittings and Materials Directory, if it is listed it can be used;

ii) telephone the Water Regulations Advisory Scheme for advice; or

iii) check with your local water undertaker.

Other cistern control valves

Inlet pipes to cisterns are required under Paragraph 16(2) **to have a servicing valve fitted immediately before connection to the cistern** to provide individual control to the float operated valve.

This rule **applies to all cisterns including water storage cisterns, feed and expansion cisterns, WC flushing cisterns and urinal flushing cisterns.**

Paragraph 16(3) requires **outlet pipes from cisterns to be fitted with servicing valves. The best position for the valve,** to comply with the Regulations, is as near to the point of connection to **the cistern** as is reasonably practical.

However, it may be permissible to fit servicing valves in an airing cupboard providing the distance from the cistern is short.

This rule **applies to all distributing pipes and cold feed pipes from cisterns** with one exception.

Cold feed pipes to primary heating circuits do not need to be fitted with a servicing valve because only small amounts of water are wasted when draining down, and it could lead to a dangerous situation if such a valve was inadvertently turned off.

What about overflow pipes and warning pipes

Water Byelaws used to be concerned with both warning pipes and overflow pipes but now Paragraph 16(4)(a) of Schedule 2 simply says that '– *every storage cistern shall be fitted with an overflow pipe, with a suitable means of warning of an impending overflow –*'.

However, to comply with the Regulations you may need in some cases to fit both an overflow pipe and a warning pipe or some other means of warning that an overflow is likely to occur. This module follows the guidance document in retaining the distinction between an overflow pipe and a warning pipe and is written accordingly.

The provision of overflow pipes and warning pipes may vary depending on the capacity of the water storage cistern, but first, perhaps you should be sure you know how a warning pipe differs from an overflow pipe.

'Overflow pipe' *means a pipe from a cistern in which water flows only when the water level in the cistern reaches a predetermined level'* (paragraph 1) It is used to discharge any overflowing water where it will not cause damage to the building, usually caused by a float operated valve that is letting by.

A **'warning pipe'** is an overflow pipe used to give warning to the owners or occupiers of a building that a cistern is overflowing and needs attention.

See also definitions.

So how does the size of the cistern affect the provision of overflow pipes and warning pipes?

To answer this, it is perhaps best to look at some diagrams of cisterns of various sizes to see clearly how warning pipes and overflow pipes may be arranged to meet the requirements of the Regulations.

(a) Small cisterns of up to 1000 litres capacity

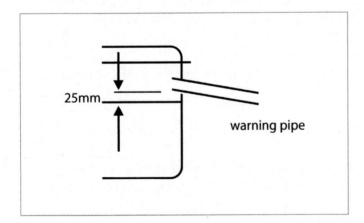

A combined overflow/warning pipe may be accepted on small cisterns with a capacity of 1000 litres or less

(b) Medium cisterns greater than 1000 litres and up to 5000 litres capacity

Cisterns with a capacity greater than 1000 litres should be capable of being inspected and cleansed without having to be wholly uncovered.

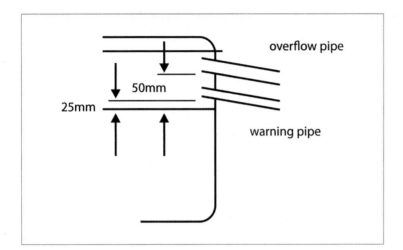

*Medium cisterns must be fitted with a **warning pipe and an overflow pipe***

(c) Large cisterns greater than 5000 litres and up to 10,000 litres capacity

These are required to be fitted with any one of a variety of warning/overflow devices which may include the following:

i)   a **warning pipe and an overflow pipe** as for medium cisterns, (see sketch of medium cistern above) or

ii)   cistern with **electrically operated warning device** which should be readily visible and show clearly when the water level is at the point of overflowing, and an overflow pipe or

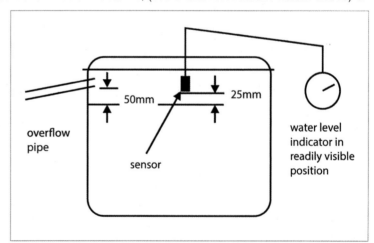

iii) a **float operated water level indicator,** and an overflow pipe

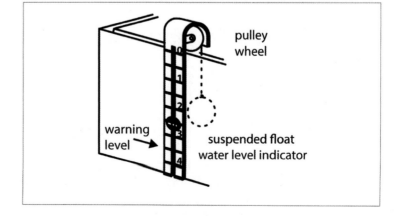

(d) Very large cisterns of more than 10,000 litres capacity

These should be fitted with one of the following:

i)   **a warning pipe and an overflow pipe** as for medium and large cisterns

ii)   **an electrically operated audible or visual alarm which clearly shows when the water level rises to within 50mm of the cistern overflowing level,** and an overflow pipe

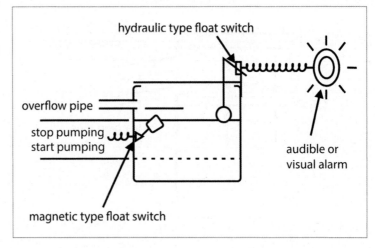

iii) **a hydraulically operated, audible or visual alarm that clearly shows when the water level rises to within 50mm of the cistern overflowing level,** and an overflow pipe

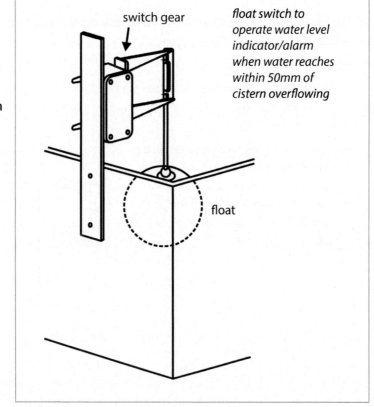

Note: the overflowing level is the point at which water begins to flow over the invert (lowest inside) of the warning pipe.

What else is important about overflow and warning pipes?

There are a number of points still to consider if your installations are to comply with the Regulations. Eight important points are set out below.

(a) In the event of the inlet control device becoming defective, the **overflow/warning pipe should be capable of removing the excess water without the inlet becoming submerged** (even in severe situations. e.g. float comes adrift).

(b) **Warning/overflow pipe to fall continuously from its cistern connection to its point of discharge.**

(c) **Warning pipes to discharge in a conspicuous position,** preferably outside the building.

(d) **Warning pipes from feed and expansion cisterns should be separate from those serving storage cisterns.**

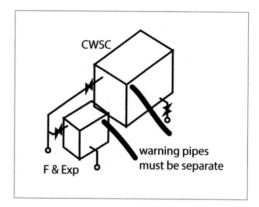

(e) **When two or more storage cisterns have a common warning pipe or overflow pipe, they should be arranged so that one cistern cannot discharge into another.**

(f) Overflow and warning pipes should be positioned so as to exclude light and screened to prevent the ingress of insects and other foreign bodies. Screen mesh size should not exceed 0.65mm (opening). See diagram page 6.

(g) Every storage cistern must be constructed and installed so that they can be easily inspected (both externally and internally) and cleansed. For ease of installation, maintenance and replacement float operated valves and other controls must be readily accessible, the minimum unobstructed clearance above a cistern is 350mm.

(h) Every cold water storage cistern should be adequately supported to avoid distortion or damage. Refer to Part G of the Building Regulations provide further information and advice.

What else does Paragraph 16 say about water storage cisterns?

Paragraph 16(4)(b) requires the **cistern to be fitted with 'a cover positioned so as to exclude light and insects.'** The diagram on page 5 illustrates this.

Paragraph 16(4)(c) says that **insulation shall be fitted to minimise freezing or undue warming.** This has been dealt with in Module 4 and does not need to be repeated now, except to say that any **insulation should include the overflow pipe or warning pipe.**

Isn't there a contamination aspect to Paragraph 16?

Yes there is! Paragraph 16(5) requires *'cisterns to be installed to minimise the risk of contamination of stored water'.* This requirement can be met to a large extent by installing a 'protected' cistern in all cases but in particular for those supplying water for domestic purposes. In other words, we should install what was formerly called a 'Byelaw 30 cistern'.

However, this paragraph goes a little further in that **cisterns are required to be** *'of an appropriate size and* **connections positioned to allow circulation and prevent areas of stagnation** *from developing'.*

There are a number of potential causes of contamination in cisterns, and the following points should be borne in mind when cisterns are designed and installed.

1. **Cisterns MUST be kept clean.**
   - They should be flushed and sterilised when installed, and
   - should be regularly cleaned and maintained when in use (every six months)

2. **Water must be kept moving with speedy replenishment of ALL stored water.**

   Cisterns should be adequately sized of course, but it is important that they are not oversized. Since the relatively recent outbreaks of legionella it is considered that many storage cisterns have been oversized in the past with a very slow throughput that may permit water in parts of the cistern to remain unchanged for quite long periods, resulting in stagnation.

3. **Outlet connections should be as low as possible** to permit sediment to pass through to taps rather than settle on the bottom of the cistern

4. **Outlet connections should be arranged so as to encourage movement of water** throughout the cistern and thus achieve regular replacement of stored water.

   At least one outlet should be positioned at the opposite end of the cistern to the inlet connection. This will also help throughput.

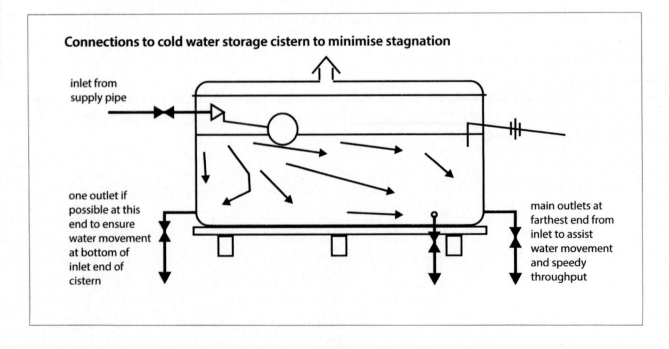

Connections to cold water storage cistern to minimise stagnation

inlet from supply pipe

one outlet if possible at this end to ensure water movement at bottom of inlet end of cistern

main outlets at farthest end from inlet to assist water movement and speedy throughput

5. **Linked cisterns** should be arranged so that they can be easily drained and cleaned, and so that accumulations of debris at the base of the cistern is discouraged.

Cisterns connected in series are generally preferred to maintain a good throughput and to reduce the risk of stagnation.

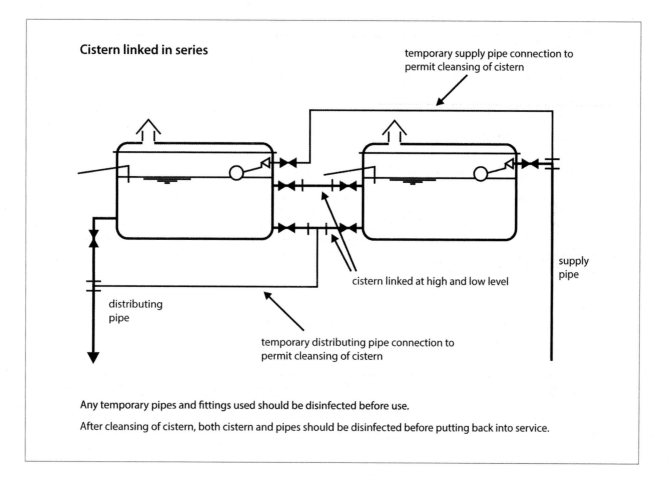

Cistern linked in series

temporary supply pipe connection to permit cleansing of cistern

cistern linked at high and low level

supply pipe

distributing pipe

temporary distributing pipe connection to permit cleansing of cistern

Any temporary pipes and fittings used should be disinfected before use.

After cleansing of cistern, both cistern and pipes should be disinfected before putting back into service.

Self-assessment questions

1. Complete the following statements outlining general requirements for storage cisterns.

    In general cisterns should be:

    a) fitted with an _____ _____ _____ to
       maintain the correct water level

    b) fitted with servicing valves on _____ and _____ pipes

    c) fitted with an overflow pipe to warn against _____ _____

    d) covered so to exclude _____ or _____

    e) insulated to prevent heat _____ and undue _____

    f) installed so that risk of contamination is _____ , and

    g) arranged so that water can circulate and _____ will not occur.

    h) supported to avoid _____ or damage that might cause them to leak

    i) readily accessible for _____ and

2. Every cold water storage cistern is required to be fitted with suitable means of shutting off the supply when the cistern is full. (usually a float operated valve). There are TWO basic requirements for this device. Give the TWO requirements.

    They must be _____ *and* _____ .

3. Name THREE types of float-operated valve that are suitable for use in water storage cisterns.

    i) _____

    ii) _____

    iii) _____

4. Give TWO positions on or near to cisterns, where servicing valves are required to be fitted.

    i) _____

    ii) _____

5. Cisterns for domestic purposes must be 'protected' against the possibility of contamination. Complete the diagram below to give the four important points of protection that should be applied to this cistern to keep it safe from contamination.

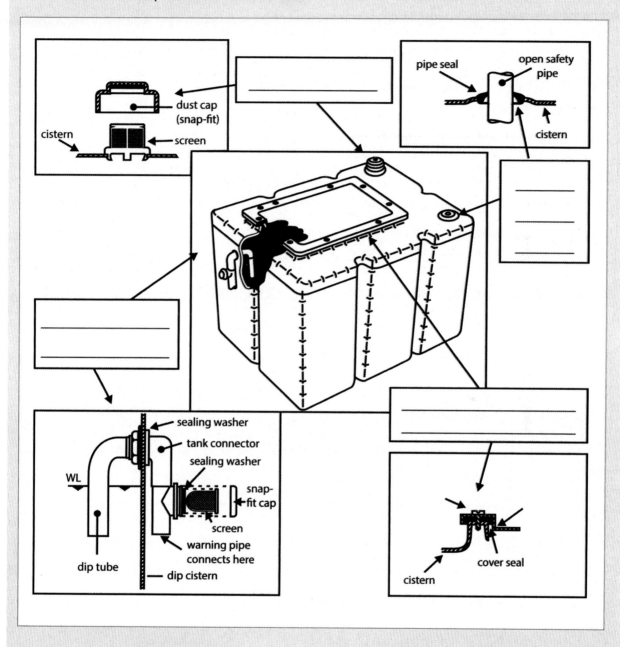

6. State the purpose of: a) a warning pipe and   b) an overflow pipe

a) _____

b) _____

_____

7. Look at the following statements and decide which is correct.
Answer Yes or No

i)   A small cistern of less than 1000 litres capacity must be fitted with
a warning pipe and no other overflow pipe   Yes ☐   No ☐

ii) A cistern of greater than 1000 litres and up to 5000 litres capacity
may be fitted with a warning pipe **or** an overflow pipe   Yes ☐   No ☐

iii) A cistern of greater than 1000 litres and up to 5000 litres capacity
must be fitted with a warning pipe **and** an overflow pipe   Yes ☐   No ☐

8. Give TWO warning devices that may be used in place of a warning pipe in a cistern of devices greater than 1000 litres and up to 5000 litres capacity.

i) _____

ii) _____

9. Complete the following statement!
Instead of a warning pipe, cisterns of more than 10,000 litres capacity may be fitted with an electrically operated alarm or a hydraulically operated alarm that clearly shows when the cistern water level rises to within:

_____   _____

10. Complete the FIVE following statements

a) *The overflow/warning pipe should be capable of removing the excess water*
   without the inlet becoming _____

b) *The warning/overflow pipe should* _____ *from*
   its cistern connection to its point of discharge

c) *Warning pipes should discharge in a* _____

d) *Warning pipes from feed and expansion cisterns are required to be* _____ *from*
   those serving storage cisterns

e) *When two or more cisterns have a common warning pipe, they should be arranged so that*

_____

11. Cisterns should have adequate support.

   Describe briefly how the support for a flexible cistern of polypropylene or GRP might differ from the support for a rigid cistern of galvanised steel or copper.

_____

_____

_____

_____

_____

Check your answers on pages 19, 20, and 21.

Summary of main points

In general **cisterns should be**:

(a) fitted with an effective inlet control device to maintain the correct water level

(b) fitted with servicing valves on inlet and outlet pipes

(c) fitted with a screened warning/overflow pipe to warn against impended overflow

(d) covered so to exclude light or insects, and

(e) insulated to prevent heat losses and heat gains

(f) installed so that risk of contamination is minimised, and

(g) arranged so that water can circulate and stagnation will not occur.

(h) supported to avoid distortion or damage that might cause them to leak

(i) readily accessible for inspection and cleansing

Every pipe supplying water to a storage cistern must be fitted with an effective adjustable shut off device that will close when the water reaches its normal level (usually a float operated valve).

Float operated valves for cisterns should comply with BS 1212: Parts 1, 2 or 3, or be WRAS approved.

Cisterns are required to be fitted with servicing valves:

– **on the supply to the cistern immediately before the float operated valve**

– **on distributing pipes as near as is practicable to the cistern connection**

Cold feed pipes to primary heating circuits do not need to be fitted with a servicing valve.

Warning pipes and overflow pipes. Suitable methods include:

(a) **Small cisterns of up to 1000 litres capacity fitted with a warning pipe and no other overflow pipe**

(b) **Medium cisterns greater than 1000 litres and up to 5000 litres capacity fitted with a warning pipe and an overflow pipe**

(c) **Large cisterns greater than 5000 litres and up to 10,000 litres capacity fitted with either:**

   i) **a warning pipe and an overflow pipe** as for medium cisterns, or

   ii) an **electrically operated warning device** which should be readily visible and show clearly when the water level is about to overflow, **and an overflow pipe, or**

   iii) a float operated water level indicator and an overflow pipe

More rules for warning/overflow pipes

(a) In the event of inlet valve failure, the **overflow/warning pipe should be capable of removing the excess water without the inlet becoming submerged**

(b) **Warning/overflow pipe to fall continuously from its cistern connection to its point of discharge**

(c) **Warning pipes to discharge in a conspicuous position,** preferably outside the building.

(d) **Warning pipes from feed and expansion cisterns to be separate from those serving storage cisterns.**

(e) **When two or more cisterns have a common warning pipe, they should be arranged so that one cistern cannot discharge into another.**

Answers to self-assessment questions

1. Complete the following statements outlining general requirements for storage cisterns.

   In general cisterns should be:

   a) *fitted with an* **effective inlet control device** *to maintain the correct water level*

   b) *fitted with servicing valves on* **inlet** *and* **outlet** *pipes*

   c) *fitted with a screened warning/overflow pipe to warn against* **impending overflow**

   d) *covered so to exclude* **light** *or* **insects**

   e) *insulated to prevent heat* **losses** *and undue* **warming**

   f) *installed so that risk of contamination is* **minimised,**

   g) *arranged so that water can circulate and* **stagnation** *will not occur.*

   h) *supported to avoid* **distortion** *or damage that might cause them to leak*

   i) *readily accessible for* **inspection** *and* **cleansing**

2. Every cold water storage cistern is required to be fitted with suitable means of shutting off the supply when the cistern is full. (usually a float operated valve) There are TWO basic requirements for this device. Give the TWO requirements.

   They must be **effective** *and* **adjustable.**

3. Name THREE types of float-operated valve that are suitable for use in water storage cisterns.

   i)   **BS 1212: Part 1 Portsmouth type**

   ii)   **BS 1212: Part 2 Diaphragm type (brass)**

   iii)   **BS 1212: Part 3 Diaphragm type (plastic)**

         or another approved float-operated valve

4. Give TWO positions on or near to cisterns, where servicing valves are required to be fitted.

   i)   **On the supply pipe, immediately before the float-operated valve.**

   ii)   **On distributing pipes, near to the cistern connection.**

5. Cisterns for domestic purposes must be 'protected' against the possibility of contamination. Complete the diagram below to give the four important points of protection that should be applied to this cistern to keep it safe from contamination.

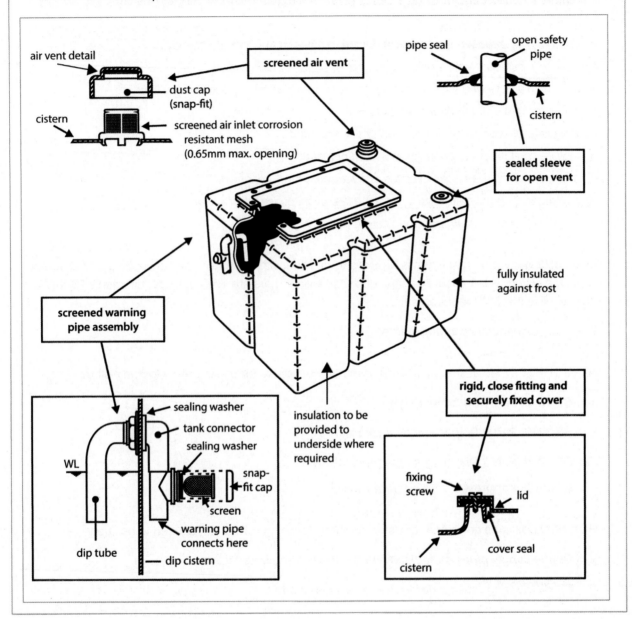

6. State the purpose of: **a)** a warning pipe, and **b)** an overflow pipe

   (a) *A warning pipe is used to give warning to the owners and occupiers of a building that the water is overflowing and needs attention.*

   (b) *An overflow pipe is used to remove from the building any water that rises above the normal water level, usually caused by a float operated valve that is letting bye.*

7. Look at the following statements and decide which is correct. Answer Yes or No.

   i) *A small cistern of less than 1000 litres capacity must be fitted with a warning pipe and no other overflow pipe*   Yes ✔   No ☐

   ii) *A cistern greater than 1000 litres and up to 5000 litres capacity must be fitted with a warning pipe **or** an overflow pipe*   Yes ☐   No ✔

   iii) *A cistern greater than 1000 litres and up to 5000 litres capacity must be fitted with a warning pipe **and** an overflow pipe*   Yes ✔   No ☐

8. Give TWO warning devices that may be used in place of a warning pipe in a cistern of devices greater than 1000 litres and up to 5000 litres capacity.

   i) *an electrically operated alarm **which should be readily visible***

   ii) *a float water level indicator which should be readily visible*

9. Complete the following statement!

   i) *Instead of a warning pipe, cisterns of more than 10,000 litres capacity may be fitted with an electrically or hydraulically operated alarm that clearly shows when the cistern water level rises to within **50mm of the cistern overflowing level.***

10. Complete the FIVE following statements

    a) *The overflow/warning pipe should be capable of removing the excess water without the inlet becoming **submerged***

    b) *The warning/overflow pipe should **fall continuously** from is cistern connection to its point of discharge*

    c) *Warning pipes should discharge in a **prominent position***

    d) *Warning pipes from feed and expansion cisterns are required to be **separate** from those serving storage cisterns*

    e) *When two or more cisterns have a common warning pipe, they should be arranged **so that one cannot discharge into another.***

11. Cisterns should have adequate support. Describe briefly how the support for a flexible cistern of polypropylene or GRP might differ from the support for a rigid cistern of galvanised steel or copper.

    Flexible cisterns need continuous support over the whole of its base area.

    Rigid cisterns e.g. galvanised steel cisterns do not need continuous support, and strategically placed timber bearers are quite adequate.

What to do next

Well: you have completed **Cold water services**

I hope you found it reasonably straightforward

Now go on to **Module 10** Hot water services

Water Industry Act 1991:

Water Supply (Water Fittings) Regulations 1999

An Open Learning Course

Module 10

Hot water services

Introduction

Much of the regulation controlling the installation of hot water services is covered by Paragraphs 17 to 24 of Schedule 2.

The requirements for hot water services are concerned primarily with the **prevention of waste of water**, but in doing that they also provide a degree of safety to the building and its occupants.

They look at expansion of hot water and the **measures required to control expansion both in vented and unvented systems**.

They deal with **safety devices for the control of water temperature in hot water storage heaters**.

The Regulations also require that **discharges** from temperature relief valves and expansion relief valves **terminate where they can readily be seen**.

Paragraph 24 prohibits the permanent connection of closed circuits to supply pipes or distributing pipes unless backflow prevention devices are in place.

What are the Requirements?

Schedule 2: Hot water services Paragraphs 17 to 24

17.-(1) *Every unvented water heater, not being an instantaneous water heater with a capacity not greater than 15 litres, and every secondary coil contained in a primary system shall:*

    (a) *be fitted with a temperature control device and either a temperature relief valve or a combined temperature and pressure relief valve;* **or**

    (b) *be capable of accommodating expansion within the secondary hot water system.*

  (2) *An expansion valve shall be fitted to ensure that the water is discharged in a correct manner in the event of a malfunction of the expansion vessel or system.*

18.   *Appropriate vent pipes, temperature control device and combined temperature pressure and relief valves shall be provided to prevent the temperature of the water within a secondary hot water system from exceeding100°C.*

19.   *Discharges from temperature relief valve, combined temperature pressure and relief valves and expansion valves shall be made in a safe and conspicuous manner.*

20.-(1) *No vent pipe from a primary circuit shall* **determine** *over a storage cistern containing wholesome water for domestic supply or for supplying water to a secondary system.*

  (2) *No vent pipe from a secondary circuit shall terminate over any combined feed and expansion cistern* **connection** *to a primary circuit.*

21.   *Every expansion cistern or expansion vessel, and every combined feed and expansion cistern connected to a primary circuit, shall be such as to accommodate any expansion water from that circuit during normal operation.*

22.-(1) *Every expansion valve, temperature relief valve or combined temperature and pressure relief valve connected to any fitting or appliance shall close automatically after a discharge of water.*

  (2) *Every expansion valve shall:*

    (a) *be fitted on the supply pipe close to the hot water vessel and without any intervening valves.*

    (b) *only discharge water when subjected to a water pressure of not less than 0.5 bar (50kPa) above the pressure to which the hot water vessel is, or is likely to be, subjected in normal operation.*

Paragraphs 23 and 24 are continued on the following page

The Requirements (continued from previous page)

> **23.-(1)** *A temperature relief valve or combined temperature and pressure relief valve shall be provided on every unvented hot water storage vessel with a capacity greater than 15 litres.*
>
> **(2)** *The valve shall:*
>   (a) *be located directly on the storage vessel in an appropriate location, and have a sufficient discharge capacity, to ensure that the temperature of the stored water does not exceed 100°C; and*
>   (b) *only discharge water at below its operating temperature when subjected to a pressure of not less than 0.5 bar (50kPa) in excess of the greater of the following:*
>   (i) *the maximum working pressure in the vessel in which it is fitted, or*
>   (ii) *the operating pressure of the expansion valve.*
>
> **(3)** *In this paragraph 'unvented hot water storage vessel' means a hot water storage vessel that does not have a vent pipe to the atmosphere.*
>
> **24.** *No supply pipe or secondary circuit shall be permanently connected to a closed circuit for filling a heating system unless it incorporates a backflow prevention device in accordance with a specification approved by the regulator for the purposes of this schedule.*

Note. You will perhaps have noticed that in the above extract from Schedule 2, three words are highlighted, one in Paragraph 17, and two in Paragraph 20.

The requirements of Paragraphs 17 and 20 are presented here as they are written in Schedule 2 of the statutory document, but there are obvious mistakes in what has been printed.

Paragraph 17 requires unvented systems to be fitted with either temperature controls or expansion controls and it is well known that **both temperature and expansion devices are essential for the unvented system to function safely and efficiently**.

In Paragraph 20(1) the word '**determine**' should read '**terminate**' and in 20(2) **connection** should read '**connecting**'. These two requirements were well documented in the past having been in water byelaws for very many years.

So! For the remainder of this manual we follow the guidance document in ignoring the errors written into Schedule 2 (even though it is law). We shall look at the intention of these requirements and interpret them as though the errors had not occurred.

What do we need to know about expansion of hot water?

Paragraph 17(1) of Schedule 2 requires **unvented water heaters to be capable of accommodating expansion of hot water within the secondary hot water system**, whilst **Paragraph 17(2) requires the provision of expansion valves** to protect the system from the possibility that the expansion vessel or system might fail.

Water heaters of 15 litres capacity or less are now required to meet the requirements of G3. G3 states *"water heaters that have a capacity of 15 litres or less that have appropriate safety devices for temperature and pressure will generally satisfy the requirement set in G3(3)"*.

In practice, this means that **two levels of control** must be applied to regulate the expansion of water in an unvented hot water system.

Firstly, it is important that **unvented hot water heaters above 15 litres capacity** are fitted with the means to **control the expansion and contraction** of water as it is heated and cooled during use. This means the unvented heater or system **must be fitted with either an expansion vessel** connected on to the supply pipe near the inlet to the heater, **or an integral expansion unit** within the water heater. These types of device are illustrated in the following diagrams.

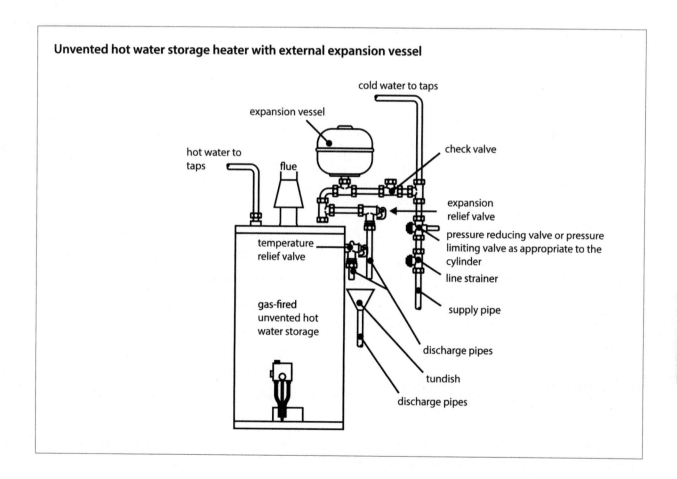

Unvented hot water storage heater with external expansion vessel

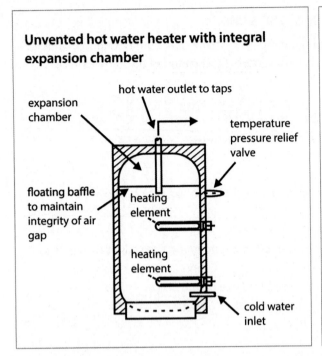

Unvented hot water heater with integral expansion chamber

hot water outlet to taps

expansion chamber

temperature pressure relief valve

floating baffle to maintain integrity of air gap

heating element

heating element

cold water inlet

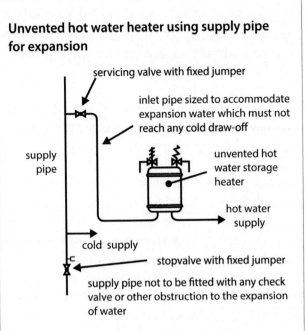

Unvented hot water heater using supply pipe for expansion

servicing valve with fixed jumper

inlet pipe sized to accommodate expansion water which must not reach any cold draw-off

supply pipe

unvented hot water storage heater

hot water supply

cold supply

stopvalve with fixed jumper

supply pipe not to be fitted with any check valve or other obstruction to the expansion of water

In recognition that water stagnation and particulate accumulation can have detrimental effects upon water quality it is recommended that any fitting on a wholesome water system which accommodates expansion or pressure surges, such as expansion vessels, pressure accumulators and surge arrestors, be installed so as to avoid localised low turnover (stagnation) leading to the formation of biofilms and/or the accumulation of particulates.

Installation, bottom fed and upright. That the connecting pipework to the fitting rises continuously and is kept to a minimum. Sized correctly for the system and designed to ensure an adequate turnover of water within the fitting.

Examples of good practice

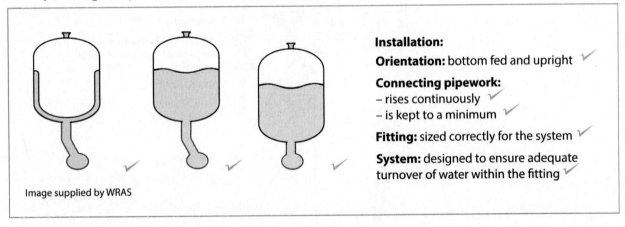

Installation:

Orientation: bottom fed and upright ✓

Connecting pipework:
– rises continuously ✓
– is kept to a minimum ✓

Fitting: sized correctly for the system ✓

System: designed to ensure adequate turnover of water within the fitting ✓

Image supplied by WRAS

To meet the requirement, expansion vessels/units will need to accommodate the equivalent of at least 4% in volume **of the total capacity of the hot water heater.**

Secondly, because expansion vessels/units are mechanical devices (and all mechanical devices have the potential to fail), Paragraph 17(2) requires that **unvented hot water heaters must be fitted with an expansion relief valve** to guard against the possible failure of the expansion vessel or unit.

<table>
<tr><td>

Paragraph 22

The expansion relief valve shall:

- close automatically after a discharge of water

- be fitted on the supply pipe close to its connection to the heater without any intervening valves, and

- only discharge when pressure rises 0.5 bar (50kPa) above the operating pressure of the cylinder

</td><td>

Expansion (pressure) relief valve

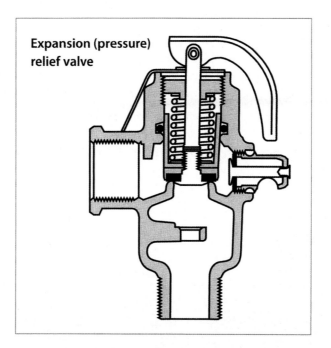

</td></tr>
</table>

Are there any rules relating to the control of temperature in hot water systems?

Yes there are! Paragraph 18 requires that the **temperature of water within a secondary hot water vessel shall NOT exceed 100°C**. To prevent this from occurring, all hot water vessels must be fitted with 'appropriate' vent pipes, temperature control devices, safety devices or dedicated cold water storage cisterns.

You will see that there is an element of safety in this requirement which was not in previous water byelaws. Safety in hot water supply has in the past been solely the concern of the Building Regulations. So! Quite a departure from tradition here!

Paragraph 18 is concerned with the prevention of hot water vessels from bursting and in particular from exploding. Hence the requirement to prevent water in hot water storage vessels from exceeding 100°C.

When water at atmospheric pressure (1 bar or 14.7 lb/sq.inch) is heated to 100°C it will simply boil and turn to steam. When turning to steam the water expands by approximately 1600 times its original volume.

1 litre of water, when boiled, equals 1600 litres of steam, approximately.

This is no problem when the water is heated in an open vessel because the steam generated will disperse quite naturally in the open air. However, when the water is contained, as it is in a hot water cylinder, the water will try to expand and change to steam, but it cannot because it is in a restricted space. As a result, the water pressure will increase unless some device is fitted to accommodate expansion and prevent damage to the cylinder.

Let's take this a stage further.

Water under pressure greater than atmospheric pressure boils at higher temperatures than 100°C depending on the pressure. For instance water at 3 bar pressure will have a boiling point of 143°C. The higher the pressure the higher the boiling point.

So to the point!

Consider a hot water cylinder that is under pressure from a supply pipe (let's say 3 bar pressure) and its temperature rises above normal boiling point to 120°C. Consider again the pressure of the expanding heated water putting the cylinder under more strain than it can take so that it splits. The loss of water through the split will cause a drop in pressure, the water will immediately boil, flash to steam, and explode with disastrous results.

The power of an exploding cylinder is like a bomb and can take a whole house apart.

(See photograph below)

Photograph of damaged building through over-heated cylinder.

What sort of vessels might possibly explode?

Well! None of them! Providing they are correctly installed and temperatures properly controlled. But! If incorrectly installed, both vented and unvented systems can be dangerous, which brings us back to the requirement of Paragraph 18, which calls for the installation of 'appropriate vent pipes and temperature devices'.

People usually think of unvented systems when mention is made of the possibility of hot water vessels exploding, through incorrectly fitted or poorly maintained safety devices. However, an open vented hot water system can be equally hazardous without proper provision for hot water to expand or for the escape of boiling water or steam if the system overheats, particularly if the cold feed pipe and vent should become frozen in cold weather.

Remember! The requirement of Paragraph 18 is to prevent the temperature of water in a secondary hot water system from reaching 100°C.

If the temperature of water in hot water storage vessels is kept below 100°C, the danger of explosion is eliminated.

How then can we prevent over heating in hot water vessels, and at the same time comply with Paragraph 18 of Schedule 2?

There are three levels of control for hot water storage installations.

1. Effective thermostatic control
2. Temperature operated thermal energy cut-off device
3. Temperature relief and heat dissipation.

1. **Effective thermostatic control** should be provided for the efficient use of heat energy. This will be in the form of a **cylinder thermostat** or an **immersion heater thermostat** to keep the temperature of stored water from rising above the normal expected day to day **temperature of 60°C to 65°C.**

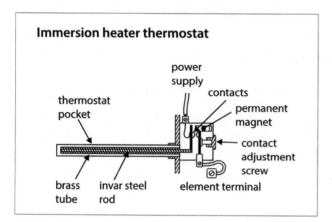

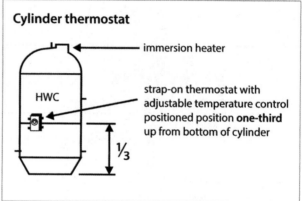

2. **Temperature operated thermal energy cut-off device.** Also known as a high energy cut-out device, this second level control is **used to shut off the energy supply at a pre-determined temperature, normally at 90°C.**

This is a type of thermostat that **must be of the non-self re-setting type and will operate only if the normal thermostatic control should fail,** and the hot store vessel begins to over heat.

For the temperature cut out to be effective it **should be fitted within the top 20% volume of the hot store vessel, or on the primary flow to the cylinder, or on the boiler.**

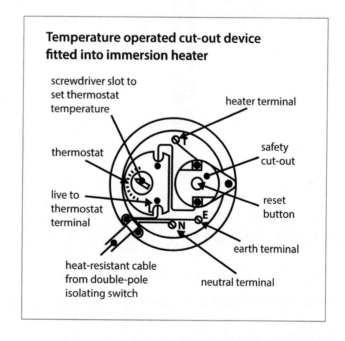

3. **Temperature relief and heat dissipation is automatically applied only if,** in the extreme case that **the other two thermostatic controls should fail to operate.**

Let's look at unvented systems first.

In unvented systems a **temperature relief valve** is used to permit overheated water to discharge safely from the hot store vessel before it can boil. It will begin to operate at about 95°C.

It is important that the water discharge from hot store vessels through temperature relief valves, and indeed from expansion relief valves, are discharged in a safe and visible manner.

Temperature relief valve

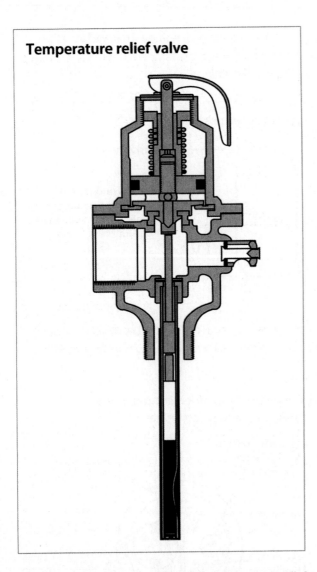

Paragraph 23

A temperature relief valve shall be provided on every unvented hot water vessel. Below 15 litres capacity water heaters will generally have appropriate safety devices for temperature and pressure included.

The temperature relief valve shall:

• be located on the storage vessel and have sufficient discharge capacity to ensure the temperature does not exceed 100°C

• only discharge water below its operating temperature if the pressure in the cylinder exceeds either:

– the maximum working pressure of the hot store vessel,

or

– the operating pressure of the expansion valve,

whichever is the greater.

The rules for discharge pipes from temperature relief valves will follow those previously set out in building regulations. and are the subject of Paragraph 19. (See page 11)

What about heat dissipation in vented hot water systems?

In open vented systems Paragraph 18 requires the use of a vent to dissipate excess heat should the hot store vessel overheat and boil. The vent pipe, perhaps termed the 'open safety vent', will provide a safe route for boiling water and steam from the top of the hot store vessel to terminate over the feed cistern.

A similar arrangement is needed on vented primary circuits, but Schedule 2 seems only to be concerned with vents to secondary hot water circuits and makes no mention of primary circuits!

The following diagram shows how the requirements of Paragraph 18 can be met in respect of the vented secondary hot water storage vessel and also shows an open safety vent to a primary circuit.

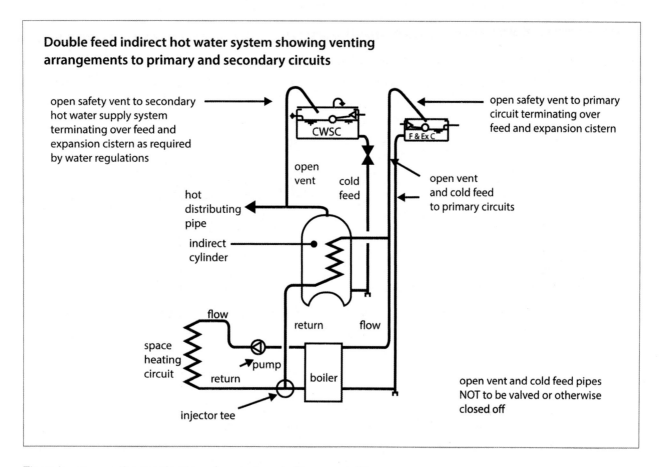

Double feed indirect hot water system showing venting arrangements to primary and secondary circuits

open safety vent to secondary hot water supply system terminating over feed and expansion cistern as required by water regulations

CWSC

open safety vent to primary circuit terminating over feed and expansion cistern

F & Ex C

open vent

cold feed

hot distributing pipe

open vent and cold feed to primary circuits

indirect cylinder

flow

space heating circuit

return

pump

return   flow

injector tee

boiler

open vent and cold feed pipes NOT to be valved or otherwise closed off

There is more on the termination of vent pipes in Paragraph 20.

Paragraph 20-(1). No vent pipe from a primary circuit shall terminate over any storage cistern containing wholesome water for domestic supply for supplying water to secondary system. Why? Quite simply because any water discharged from the primary circuit is likely to contaminate the stored water supplying the secondary system.

Paragraph 20(2). No vent pipes from a secondary circuit shall terminate over any feed and expansion cistern connected to primary circuit. Why? Because the feed and expansion cistern is too small to accommodate the water from a secondary circuit and would overflow to cause waste of water.

We discussed earlier, expansion of heated water in unvented systems.

But! What about expansion in vented systems?

This is dealt with in Paragraph 21.

It is just as important to **control expansion in vented secondary circuits** as it is in unvented systems. Unless expansion is allowed to take place there is a danger of the hot water cylinder being damaged and perhaps bursting.

As water in the hot store vessel is heated, it expands and the cooler water in the bottom of the cylinder is displaced, to return back into the cold feed to the feed cistern that supplies the vessel.

The amount of expansion to be allowed for is an amount equal to 4% of the volume of water contained in the secondary hot water system.

For example, in the case of a 100 litre hot water cylinder the expansion allowance would be an additional 4 litres.

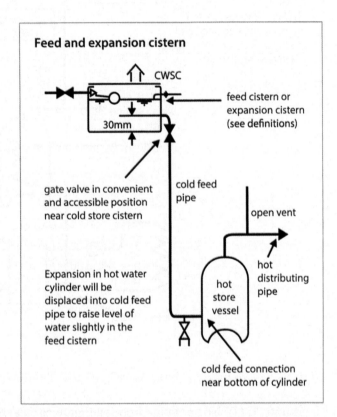

Feed and expansion cistern

Feed and expansion cisterns must be able to accommodate any expansion of water in a primary circuit. That means the water level in the cistern should be low enough to allow room for the expansion without the water level rising above a point 25mm below the overflowing level of the cistern. The overflowing level being at the lowest inside part of the overflow/warning pipe.

The amount of expansion to be allowed for is an amount equal to 4% of the volume of water contained in the primary heating circuit.

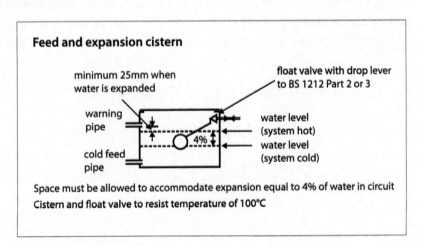

Feed and expansion cistern

What are the rules relating to discharges from temperature relief valves and expansion/pressure relief valves?

Paragraph 19 of Schedule 2 says that:

Discharges from temperature relief valves, combined temperature pressure relief valves and expansion relief valves must be made in a safe and conspicuous manner:

They must discharge in:

i)   **a readily visible position** where any discharge is seen and dealt with before a disaster occurs; and

ii)   **a safe place** where it cannot cause injury to occupants of the building or other third persons nearby.

How these two points can be achieved in practice is illustrated in the diagrams below.

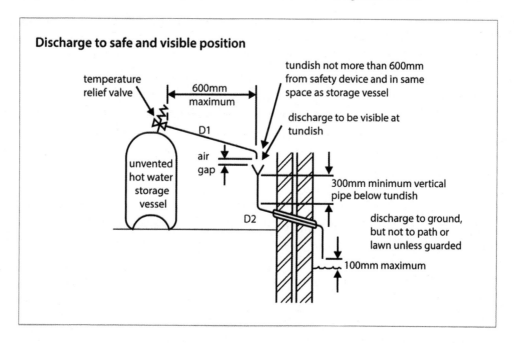

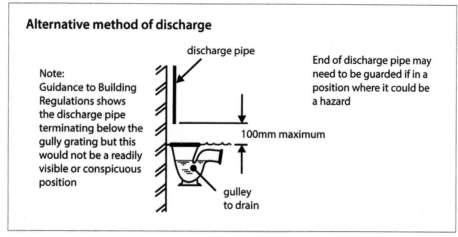

Finally Paragraph 24 – connections to primary circuits

This is quite a simple requirement, that '**no supply pipe or a distributing pipe shall be permanently connected to a closed circuit for filling a heating system unless it incorporates a backflow device in accordance with a specification approved by the Regulator for the purposes of this Schedule**'.

Primary and other closed circuits have to be initially filled with water and may require additional 'topping-up' at intervals during use. Primary circuits may contain additives and the water can be heavily contaminated, therefore they are not to be permanently connected to any supply pipe without an adequate backflow prevention device.

Where a connection is made to a supply pipe, or a distributing pipe in some instances, for supplying water for filling or replenishing water in a closed circuit, such as a hot water primary circuit and/or a space heating system it is essential that:

(a) there is no backflow of water at any time, from the primary circuit into the water supply; and

(b) the water supply is disconnected, or vented to atmosphere during the periods between filling and subsequent replenishing of the water in the primary circuit.

Under normal operating conditions, the pressure in a primary heating circuit is less than in the pipe supplying water to the circuit. However in the event of a malfunction of an expansion valve or pressure relief valve in the primary circuit, pressure may rise above the pressure in the supply pipe. In such an instance, a mechanical backflow prevention device could be damaged and cease to function. If there is no discontinuity or venting to atmosphere and, as frequently happens, the valve controlling the water supply has been left in the open position, fluid from the primary circuit may return to the supply pipe.

It is therefore essential that when the filling or the replenishing of a primary circuit is completed, there shall be a discontinuity at the point of connection, or the type of backflow prevention device installed shall be of a type that allows any fluid resulting from excess pressure in the primary circuit to discharge to waste.

The type of backflow prevention device required should be suitable for a fluid category 3 risk in the case of a house or for fluid category 4 risk for installations in premises other than a house. It is however essential that there is a discontinuity in the connecting pipework or a backflow prevention arrangement is used in which any fluid resulting from backflow from the primary circuit is discharged to waste.

To avoid control valves being tampered with and left in an open position it is recommended that all control valves used in connection with filling loops should be lockshield type valves with a loose key.

A satisfactory method of filling or replenishing a primary circuit in a house is show in the first diagram below, where the temporary connecting pipe is completely disconnected after filling or replenishment.

Another method that is considered acceptable for fluid category 3 risk in a house is the installation of Type CA: 'Non-verifiable disconnector with different pressure zones' backflow prevention device (see second diagram below).

In other than a house, where backflow protection against a fluid category 4 risk is required, Type BA: 'Verifiable backflow preventer with reduced pressure zone' backflow preventer (RPZ valve), with a strainer on the inlet, could be used (see third diagram below).

Unless using an approved backflow prevention device which is permitted to operate continuously, methods of supplying feed water to a closed circuit are to be manually operated and are only to be used when make up water is required.

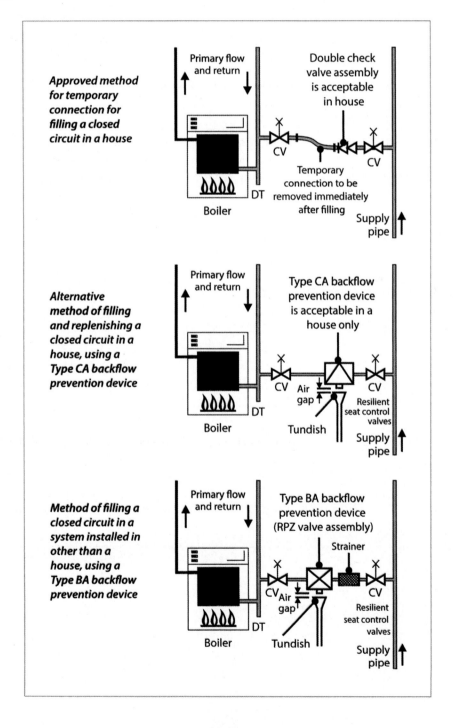

Self-assessment questions

1. Paragraph 17(1) of Schedule 2 requires unvented water heaters to be capable of accommodating expansion of hot water. Use the following Yes/No boxes to indicate whether the following statements are correct or incorrect!

   a) The requirement applies to heaters of 15 litres capacity or more   Yes ☐   No ☐

   b) The requirement applies to heaters of more than 15 litres capacity   Yes ☐   No ☐

   c) The requirement applies equally to open vented systems   Yes ☐   No ☐

2. Unvented hot water heaters may be fitted with either an expansion vessel external to the heater or an integral expansion unit.

   a) Give the approximate amount of expansion water that must be accommodated

   _____

   b) Name the device that should be fitted as a back-up in case the expansion vessel/unit fails

   _____

3. Paragraph 22 gives THREE rules for the provision of expansion relief valves. State:

   a) what should it do following a discharge of water;

   b) where it should be positioned; and

   c) what Paragraph 22 says about its operating pressure

   a) _____

   b) _____

   c) _____

4. Complete the following sentences relating to the temperature requirements for hot water systems:

   a) The temperature of a secondary hot water system shall not exceed _____

   b) The THREE levels of control required to prevent overheating of unvented hot water storage vessels are:

   1. _____

   2. _____

   3. _____

5. State the best location for the following devices to be effective:

    a) A strap-on cylinder thermostat.

    _____

    b) thermal cut-out device fitted to an unvented cylinder.

    _____

6. State what should occur if a cylinder thermostat or an immersion heater thermostat in an unvented hot water heater should fail and the water temperature rise uncontrolled.

    _____

7. State the conditions that might cause a thermal relief valve to operate.

    _____

8. Complete the following sentences

    A temperature relief valve shall be provided on every unvented hot water storage vessel of more than

    _____

    The temperature relief valve shall be located on the storage vessel to ensure the temperature does not

    _____

    The temperature relief valve shall only discharge water below its operating temperature if the pressure in the cylinder exceeds either:

    – the _____ of the hot store vessel, or

    – the _____ of the expansion valve,

    – whichever is the _____

9. State the devices or arrangements that are required by Paragraph 18 to prevent the temperature of hot water in a vented hot water storage cylinder from reaching 100°C.

    1. The device used to control the temperature of the water is:

    _____

    2. To dissipate boiling water and steam if the cylinder should overheat the device/arrangement used is:

    _____

10. There are TWO rules for the termination of vents from primary and secondary hot water circuits. Give the TWO rules by completing the following sentences.

a) A vent from a primary hot water circuit shall not

_____

b) A vent from a secondary hot water system shall not

_____

11. Expansion in a vented primary hot water circuit is accommodated in a feed and expansion cistern via the cold feed pipe.

Complete the annotations on the diagram to show the F & Ex cistern requirements.

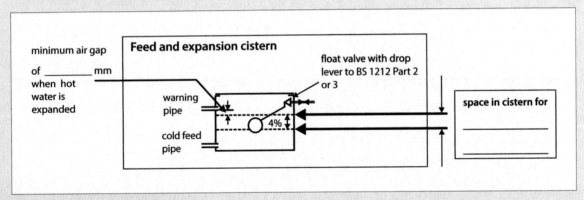

12. There are TWO rules for the discharge of temperature relief valves and expansion relief valves. (Paragraph 19) Give the TWO rules below.

i) _____

ii) _____

13. Complete the following statement relating to sealed primary circuits.

No supply pipe or distributing pipe shall be permanently connected to a closed circuit for filling a heating system unless it incorporates

_____

Check your answers on pages 20, 21 and 22.

Summary of main points

The main purpose of the hot water services Paragraphs 17 to 24 of Schedule 2 is to **prevent waste of water** and **provide some degree of safety in the control of expansion and temperature in hot water storage systems.**

Paragraph 17(1) requires **unvented hot water storage heaters of more than 15 litres capacity to be capable of accommodating expansion of hot water within the secondary hot water system, and Paragraph 17(2) requires the provision of expansion valves** to protect the system from the possibility that the expansion vessel or system might fail.

Unvented heater/systems **must be fitted with either**:

- **an expansion vessel** connected to the supply pipe near the inlet to the heater which should preferably be of the 'flow through' type; or
- **an integral expansion unit** within the water heater.

Expansion vessels/units need to be able **to accommodate expansion of at least 4% of the total capacity of the heater.**

Paragraph 18 requires that the **temperature of water within a secondary hot water vessel shall NOT exceed 100°C**. To prevent this from occurring, all hot water vessels must be fitted with 'appropriate' vent pipes, temperature control devices, safety devices or dedicated cold water storage cisterns

There are **three levels of control** for hot water storage installations.

1. **Effective thermostatic control** to maintain normal temperatures in the hot water vessel. (e.g. 60°C to 65°C)
2. **Temperature operated thermal energy cut-off device** to turn off the energy supply at a pre-determined temperature (about 90°C), if the thermostat should fail.
3. **Temperature relief and heat dissipation** is used to guard against failure of the other two controls and prevent the water from reaching a temperature of 100°C.

   In unvented systems a temperature relief valve will operate at about 95°C to **permit hot water to discharge safely** if the hot store vessel should overheat.

   In open vented systems Paragraph 18 requires the use of a vent to dissipate excess heat should the hot store vessel overheat and boil.

Paragraph 20(1). **No vent pipe from a primary circuit shall terminate over any storage cistern containing wholesome water for domestic supply or for supplying water to a secondary system.**

Paragraph 20(2). **No vent pipe from a secondary circuit shall terminate over any feed and expansion cistern connected to a primary circuit.**

Paragraph 21. **Feed and expansion cisterns must be able to accommodate any expansion of water in a primary circuit.** (At least 4% in volume of the total capacity of the primary heating circuit.)

Any discharge from temperature relief valves and expansion/pressure relief valves must be readily visible and must discharge to a safe place.

Closed primary circuits shall not be directly and permanently connected to supply pipes unless they incorporate an approved backflow prevention arrangement.

Answers to self-assessment questions

1. Paragraph 17(1) of Schedule 2 requires unvented water heaters to be capable of accommodating expansion of hot water. Use the following Yes/No boxes to indicate whether the following statements are correct or incorrect!

   a) The requirement applies to heaters of 15 litres capacity or more   Yes ☐   No ✔

   b) The requirement applies to heaters of more than 15 litres capacity   Yes ✔   No ☐

   c) The requirement applies equally to open vented systems   Yes ☐   No ✔

2. Unvented hot water heaters may be fitted with either an expansion vessel external to the heater or an integral expansion unit.

   a) Give the approximate amount of expansion water that must be accommodated

      4% of the capacity of the hot store vessel

   b) Name the device that should be fitted as a back-up in case the expansion vessel or unit fails

      Expansion relief valve

3. Paragraph 22 gives THREE rules for the provision of an expansion relief valve. State:

   a) what should it do following a discharge of water;

   b) where it should be positioned; and

   c) what Paragraph 22 says about its operating pressure

      a) **It shall close automatically** after a discharge of water

      b) **It shall be fitted on the supply pipe close to its connection to the heater, without any intervening valves**

      c) **It shall only discharge when pressure rises 0.5 bar (50kPa) above the operating pressure of the cylinder**

4. Complete the following sentences relating to the temperature requirements for hot water systems:

   a) The temperature of a secondary hot water system shall not exceed **100°C**

   b) The THREE levels of control required to prevent overheating of unvented hot water storage vessels are:

      1. **Thermostatic control**

      2. **Temperature operated thermal cut-off device**

      3. **Temperature relief and heat dissipation**

5. State the best location for the following devices to be effective:

   a) A strap-on cylinder thermostat

   one third up from the base of the cylinder

   b) thermal cut-out device fitted to an unvented cylinder

   In the top 20% volume of the cylinder

6. State what should occur if a cylinder thermostat or an immersion heater thermostat in an unvented hot water heater should fail and the water temperature rise uncontrolled.

   The thermal cut-out should operate at about 90°C

7. State the conditions that might cause a thermal relief valve to operate.

   The temperature relief valve is designed to operate if **both the cylinder thermostat and the thermal cut-out device should fail.**

8. Complete the following sentences.

   A temperature relief valve shall be provided on every unvented hot water storage vessel of more than **15 litres capacity**

   The temperature relief valve shall be located on the storage vessel to ensure the temperature does not **exceed 100°C**

   The temperature relief valve shall only discharge water below its operating temperature if the pressure in the cylinder exceeds either:
   – the **maximum working pressure** of the hot store vessel, or
   – the **operating pressure** of the expansion valve,
   – whichever is the **greater**.

9. State the devices or arrangements that are required by Paragraph 18 to prevent the temperature of hot water in a vented hot water storage cylinder from reaching 100°C.

   1. The device used to control the temperature of the water is:

   a cylinder thermostat (or immersion heater thermostat)

   2. To dissipate boiling water and steam if the cylinder should over-heat the device/arrangement used is:

   An open safety vent

10. There are TWO rules for the termination of vents from primary and secondary hot water circuits. Give the TWO rules by completing the following sentences.

 a) *A vent from a primary hot water circuit* **shall not terminate over a storage cistern supplying a secondary hot water system**

 b) *A vent from a secondary hot water system* **shall not terminate over any feed and expansion cistern**

11. Expansion in a vented secondary hot water storage vessel is accommodated in the feed and expansion cistern via the cold feed pipe.

 Complete the annotations on the diagram to show the F & Ex cistern requirements.

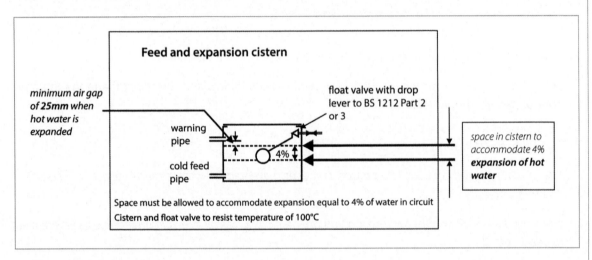

There are TWO rules set out in Requirement 21 for the discharge of temperature relief valves and expansion relief valves. State the TWO rules:

i)   **they must be readily visible, and**

ii) **they must terminate in a safe place**

12. Complete the following statement relating to sealed primary circuits (Paragraph 24).

 No supply pipe or distributing pipe shall be permanently connected to a closed circuit for filling a heating system unless it incorporates

 an appropriate backflow prevention device

What to do next

Good!

That is the hot water services manual completed. Not too bad was it?

Now go on to **Manual 11** WC flushing devices and urinals

Water Industry Act 1991:

Water Supply (Water Fittings) Regulations 1999

An Open Learning Course

Module 11

WCs, flushing devices and urinals

Introduction

This module looks at Paragraph 25 of Schedule 2 and is concerned primarily with the **conservation of water**. It deals with water closets and urinals, and in particular with the flushing arrangements.

Water Byelaws have been written in part to ensure that conservation of water is considered when installing water using appliances. For instance, WCs and urinals are known to use large quantities of water, the WC accounting for about a quarter of the water used in dwellings.

The Regulations follow previous Byelaws in the control of flushing capacities of WC cisterns to significantly reduce the quantity of water consumed.

New provisions now reduce the standard flushing volume for washdown WCs from 7.5 litres **to a maximum of 6 litres**.

The dual-flush cistern, which was discontinued by the 1986 Byelaws **is back** in fashion with a maximum six litre flush with the lesser volume a maximum of two thirds of the greater volume – i.e. four litres if the greater volume is six litres, as a means to further conserve our precious water resources, albeit with different dual flush devices.

A Regulators' Specification for the Performance of WC Suites is written to complement the Byelaws, and will require stringent standards to be met in the manufacture of WC suites. It is said, that these new provisions will give more efficient and effective flushing, permit innovative flushing mechanisms and bring us more closely into line with current European practices! The Specification was effective from **1 January 2001**, at which time the **flushing of WC cisterns by both siphonic and non-siphonic methods** will be permitted.

Modulely operated flushing valves are to be permitted for WC pans and urinals, but not in a house. In premises other than a house, these are permitted to be supplied from a supply pipe or distributing pipe provided suitable backflow prevention devices are in place.

What is the Requirement?

Schedule 2: Paragraph 25

WCs, flushing devices and urinals

25.-(1) *Subject to the following provisions of this paragraph:*

(a) *Every water closet pan shall be supplied with water from a flushing cistern, pressure flushing cistern or pressure flushing valve, and be so made and installed that after normal use its contents can be cleared effectively by a single flush of water, or, where designed to receive flushes of different volumes, by the largest of those flushes.*

(b) *No pressure flushing valve shall be installed:*

    (i) *in a house, or*

    (ii) *in any building not being a house where a minimum flow rate of 1.2 litres per second cannot be achieved at the appliance;*

(c) *Where a pressure flushing valve is connected to a supply pipe or distributing pipe, the flushing arrangement shall incorporate a backflow prevention device consisting of a permanently vented pipe interrupter located not less than 300mm above the spillover level of the WC pan or urinal;*

(d) *No flushing device installed for use with a WC pan shall give a flush exceeding 6 litres;*

(e) *No flushing device designed to give flushes of different volumes shall have a lesser flush exceeding two thirds of the largest flush volume;*

(f) *Every flushing cistern other than a pressure flushing cistern, shall be clearly marked internally with an indelible line to show the intended volume of the flush, together with an indication of that volume;*

(g) *A flushing device designed to give flushes of different volume:*

    (i) *shall have a readily discernible method of actuating the flush at different volumes; and*

    (ii) *have instructions, clearly and permanently marked on the cistern or displayed nearby, for operating it to obtain the different volumes of flush;*

(h) *Every flushing cistern, not being a pressure flushing cistern or a urinal flushing cistern, shall be fitted with a warning pipe or with a no less effective device;*

(i) *Every urinal that is cleared by water after use shall be supplied with water from a flushing device which:*

    (i) *in the case of a flushing cistern, is filled at a rate suitable for the installation;*

    (ii) *in all cases, is designed or adapted to supply no more water than is necessary for effective flow over the internal surface of the urinal and for replacement of the fluid in the trap: and*

(j) *Except in the case of a urinal which is flushed modulely, or which is flushed automatically by electronic means after use, every pipe which supplies water to a flushing cistern or trough used for flushing a urinal shall be fitted with an isolating valve controlled by a time switch and a lockable isolating valve, or with some other equally effective automatic device for regulating the periods during which the cistern may fill.*

Schedule 2 Paragraph 25 (continued from previous page)

25.-(2) *Every water closet, and every flushing device designed for use with a water closet, shall comply with a specification approved by the regulator for the purpose of this Schedule.*

(3) *The requirements of sub-paragraphs (1) and (2) do not apply where faeces or urine are disposed of through an appliance that does not solely use fluid to remove the contents.*

(4) *The requirement in sub-paragraph (1)(i) shall be deemed to be satisfied:*

    (a) in the case of an automatically operated flushing cistern servicing urinals which is filled at a rate not exceeding:

        (i) 10 litres per hour for a cistern serving a single urinal;

        (ii) 7.5 litres per hour per urinal bowl or stall, or as the case may be, for each 700mm width of urinal slab, for a cistern serving two or more urinals

    (b) in the case of modulely or automatically operated pressure flushing valve used for flushing urinals which delivers not more than 1.5 litres per bowl or position each time the device is operated.

(5) *Until 1st January 2001 Paragraphs (1)(a) and (d) shall have effect as if they provided as follows:*

    "(a) every water closet pan shall be supplied with water from a flushing cistern or trough of the valveless type which incorporates siphonic apparatus;"

    "(b) no flushing device installed for use with a WC pan shall give a flush exceeding 7.5 litres;".

(6) *Notwithstanding sub-paragraph (1)(d) a flushing cistern installed before 1st July 1999 may be replaced by a cistern which delivers a similar volume and which may be either single flush or dual flush; but a single flush cistern may not be so replaced by a dual flush cistern.*

(7) *In this paragraph:*

"pressure flushing cistern" means a WC flushing device that utilises the pressure of water within the cistern supply pipe to compress air and increase the pressure of water available for flashing a WC pan;

"pressure flushing valve" means a self closing valve supplied with water directly from a supply pipe or a distributing pipe which when activated will discharge a pre-determined flush volume;

"trap" means a pipe fitting, or part of a sanitary appliance, that retains a liquid to prevent the passage of air; and

"warning pipe" means an overflow pipe whose outlet is located where the discharge of water can readily be seen.

Note: *The period from 1 July 1999 to 1 January 2001 was transitional and allowed the installation of either 7.5 or 6 litre cisterns*

Water closets

What method of flushing is permitted for WC pans?

The Water Supply (Water Fittings) Regulations 1999 make new provisions for flushing WC suites. Flushing methods are now allowed that were not previously permitted under Water Byelaws.

WC suites have to be rigorously tested, from 1 January 2001, to ensure they meet the strict performance criteria set out in a 'Regulators' Specification for WC performance'. WC suites are not permitted unless they comply with the Regulators' Specification. [Paragraph 25(2)]

These provisions encourage the economy of water by reducing flush volumes and aim to bring us into line with European practices by introducing new, (to this country) innovative types of flushing arrangement.

Paragraph 25(a) of Schedule 2, says that **WCs may be flushed using a flushing cistern, a pressure flushing cistern or a pressure flushing valve**.

When flushed, the **WC flushing device must clear the contents of the bowl effectively using a single flush** of water.

Cisterns may be arranged to provide for **single** or **dual flushing action**.

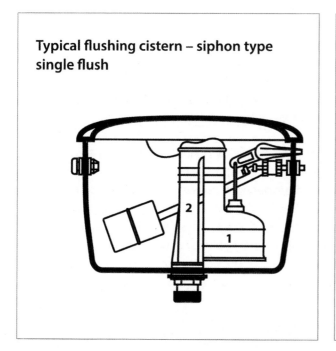

Typical flushing cistern – siphon type single flush

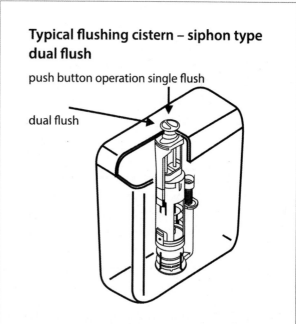

Typical flushing cistern – siphon type dual flush

push button operation single flush

dual flush

What provisions are made for flushing valves?

WCs may be flushed using a modulely operated pressure flushing valve, but only:

- in **premises other than a house** [Paragraph 25(b)(i)]
- where a minimum **flow rate of 1.2 litres per second can be achieved at the appliance** [Paragraph 25(b)(ii)]

Pressure flushing valves may be supplied from a supply pipe or from a distributing pipe [Paragraph 25(c)]

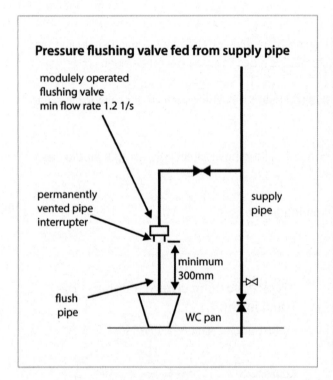

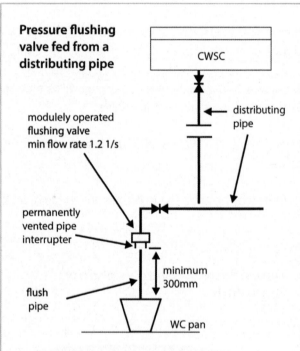

Paragraph 25(c) also provides for backflow protection for WCs supplied through pressure flushing valves.

Pressure flushing valves connected to a supply pipe or a distributing pipe are required to be fitted with a permanently vented pipe interrupter for backflow protection from the WC pan. The valve allows a pre-determined quantity to be flushed. The flushing can be initiated manually or automatically – providing the maximum consumption rates are observed. As you will recall from Module 8, a WC pan is a fluid category 5 backflow risk.

The pipe interrupter should be located:

- **at the outlet of the flushing valve;**
- **not less than 300mm above the spillover level of the pan; and**
- **there should be no tap, valve or restriction to the flow at its outlet.**

Every dual flush cistern or valve should:

– have a **clearly identifiable operating mechanism**
  e.g. separate push buttons for long and short flush, and

– have **instructions, clearly and permanently marked on the cistern** or displayed nearby, for operating it to obtain the full flush or the reduced flush

---

Instructions clearly and permanently marked on or near cistern

**press large
button for
full flush**

**press small
button for
lesser flush**

What are the permitted discharge volumes for WCs?

Paragraph 25(d) is quite specific 'no flushing device for use with a WC pan shall give a single flush of more than six litres', and:

Paragraph 25(e) says 'no flushing device designed to give a flush of different volumes shall have a lesser flush exceeding two thirds of the largest flush volume'.

So! In practice maximum permitted flushing volumes from WCs are:

– for **single flush cisterns**   six litres

– for **dual flush cisterns**   – **full flush**   six litres
                                    – **reduced flush**   four litres maximum

Are there any exceptions to the above requirements?

Paragraph 25(5) exempts flushing cisterns installed before 1 January 2001, which is the date at which the Regulators' Specification for WC Suite Performance came into force.

Until that date you could generally apply the requirements of the previous water byelaws. However, from 1 July 1999 dual-flush cisterns are permitted with a maximum six litre flush.

What about replacement of existing cisterns?

Paragraphs 25(6) similarly exempts flushing cisterns installed before 1 July 1999. This means that cisterns that were installed lawfully under Water Byelaws may be replaced by a cistern which delivers a similar volume to the original one.

In your work you will come across a variety of WC suites that have been fitted into premises over a period of many years. Flushing cisterns will range in capacity from the new 6 litre flush through to 7.5 litres, 9 litres and occasionally even larger where cisterns are much older. The range of cisterns you come across will include both single flush and dual flush mechanisms.

The important thing about replacing WC suites is that both pan and cistern are designed and manufactured as one unit.

So! When fitting either **a replacement cistern or pan, it must be of the same design as the original**, otherwise it is unlikely that they will operate satisfactorily.

Alternatively, both cistern and pan should be changed for a new suite.

The following table gives an outline of permitted flushing arrangements before and after 1 January 2001.

Maximum permitted capacity of WC cisterns			
Type of appliance	Use of appliance	Maximum permitted volume	Dates of application and restrictions
single flush only	domestic and non-domestic	7.5 litres	Until 1 July 1999 Flushing by cistern with siphonic device only
single flush and dual flush	domestic and non-domestic	7.5 litres 5 litres	Until 1 January 2001 Flushing by cistern with siphonic device only
single flush and dual flush	domestic only	6 litres 6/4 litres	From 1 January 2001 Flushing by cistern only Siphonic and non-siphonic devices permitted
single flush and dual flush	non-domestic	6 litres 6/4 litres	From 1 January 2001 Flushing by cistern or pressure flushing valve Siphonic and non-siphonic devices permitted

What are the rules for fitting warning pipes to WC flushing cisterns?

Paragraph 25(1)(h) requires **every flushing cistern**, other than a pressure flushing cistern, or a urinal flushing cistern, **to be fitted with a warning pipe** or some no less effective device.

As with all warning pipes, (you probably think of them as overflow pipes) some of the rules are the same as previous Byelaw Requirements, and are as follows:

– When **WC flushing cisterns** or troughs are fitted with a warning pipe, this pipe must have a minimum diameter of 19mm, or has to be at least one size larger than the inlet pipe, whichever is the greater.

– **The warning pipe** which terminates to atmosphere outside the building must discharge where it will be readily seen which means in a position where the property owner will notice it and repair it if it overflows.

– A '**no less effective device**' is an alternative to a warning pipe but which still provides a warning of a failure of the water inlet valve to a WC cistern.

The Water Supply Industry considers that WCs that have an **internal overflow** discharging into the WC pan shall be deemed to meet the requirement of the Byelaws in that the **internal overflow** will be regarded as a no less effective device (in place of a warning pipe). A warning pipe may also discharge directly into the flush pipe (without a tundish) as this may be considered equivalent to an **internal overflow**. These two interpretations are conditional upon measures to reduce the likelihood of the **internal overflow** being used. For compact inlet valves (e.g. those manufactured to BS 1212: Part 4) the provision of a gauze strainer incorporated in, or fitted upstream of, the float-operated inlet valve to trap debris (swarf, etc) which might cause premature failure of the valve, is required. Manufacturers are encouraged by the Water Supply Industry to supply, as a unit, the cistern with a valve and a strainer.

The Water Supply Industry considers the following to be no less effective devices:

(a) a visible warning, for example, a tundish, sight glass, mechanical signal or an electrically operated device such as an indicator lamp;

(b) an audible signal;

(c) a mechanical device which enables the flush, thereby indicating to the user that there is a fault condition in the WC flushing system;

(d) a device that detects when the water level rises above the maximum operating level and closes the water supply to the float operated valve.

And what about common warning pipes to ranges of WCs?

Individual warning pipes may discharge into a common warning pipe providing:

(a) the **common pipe serves WCs only** and they are **all at the same level,** and

(b) the **individual warning pipes discharge into a tundish** which is **readily visible**.

(See diagram overleaf)

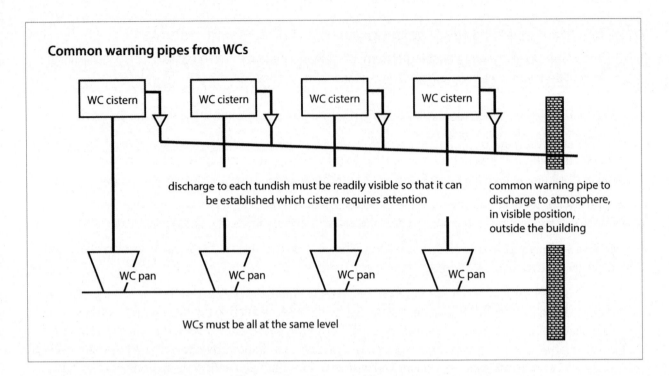

Common warning pipes from WCs

discharge to each tundish must be readily visible so that it can be established which cistern requires attention

common warning pipe to discharge to atmosphere, in visible position, outside the building

WCs must be all at the same level

Can the overflow pipe be discharged into the WC pan?

Yes. An internal overflow discharging into the WC pan can be used, provided that a ***Type AG Air Gap*** is provided between the lowest level of the outlet of the float valve and the discharge from the overflow pipe.

From January 2001 manufacturers started to produce only six litre cisterns which have some type of internal overflow therefore the arrangements shown above and below will become obsolete though they will remain legal.

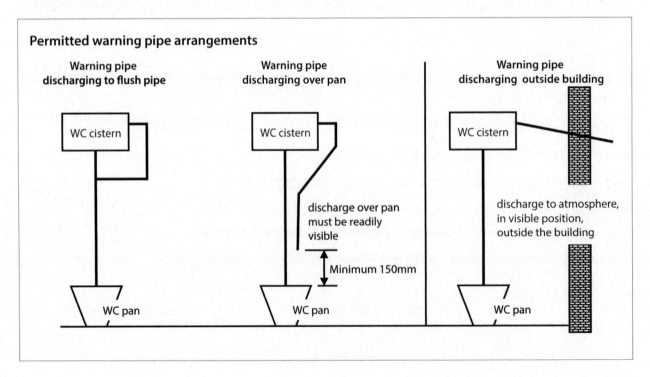

Permitted warning pipe arrangements

Warning pipe discharging to flush pipe

Warning pipe discharging over pan

Warning pipe discharging outside building

discharge over pan must be readily visible

Minimum 150mm

discharge to atmosphere, in visible position, outside the building

Flushing of urinals

How may urinals be flushed?

Devices used for flushing urinals should be 'designed or adapted to supply no more water than is necessary' for the flush to effectively clear the urinal and replace the trap seal. [Paragraph 25(i)]

There are a number of ways of operating urinals that are acceptable under the Regulations.

- by **flushing cistern** operated – **modulely,** or
                                      – **automatically**

- by **flushing valve** operated   – **modulely,** or
                                      – **automatically**

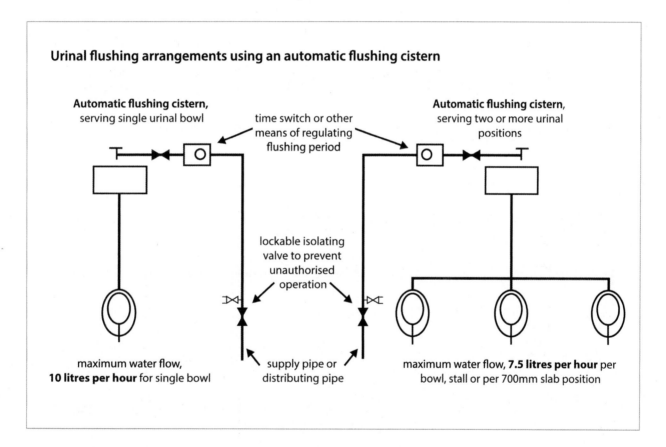

Urinal flushing arrangements using an automatic flushing cistern

Automatic flushing cistern, serving single urinal bowl

time switch or other means of regulating flushing period

Automatic flushing cistern, serving two or more urinal positions

lockable isolating valve to prevent unauthorised operation

maximum water flow, **10 litres per hour** for single bowl

supply pipe or distributing pipe

maximum water flow, **7.5 litres per hour** per bowl, stall or per 700mm slab position

How much water should a urinal use when it flushes?

Paragraph 25(4) sets out maximum volumes of water permitted for flushing.

Flushing cisterns operated automatically, require (G25-12):

(a) **10 litres per hour for a single urinal bowl, or stall;**

(b) **7.5 litres per hour, per urinal position, for a cistern serving two or more urinal bowls, stalls or per 700mm slab positions.**

Modulely operated (chain pull or push button) cisterns to single bowl urinals are required to deliver not more than **1.5 litres each time the cistern is operated.**

Pressure flushing valves operated either modulely or automatically should deliver not more than **1.5 litres each time the valve is operated.**

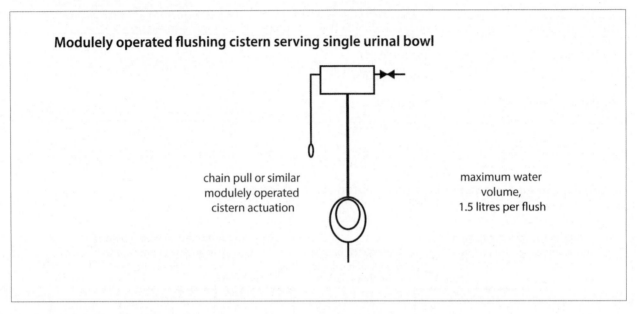

Modulely operated flushing cistern serving single urinal bowl

chain pull or similar
modulely operated
cistern actuation

maximum water
volume,
1.5 litres per flush

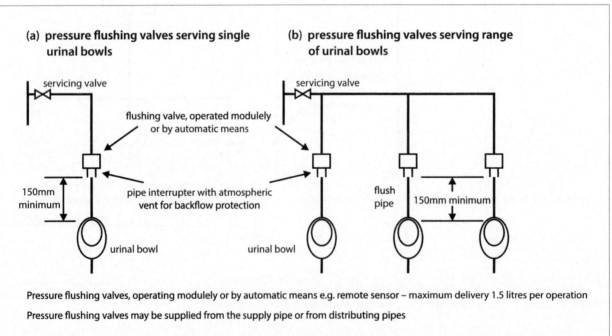

(a) pressure flushing valves serving single urinal bowls

(b) pressure flushing valves serving range of urinal bowls

servicing valve

servicing valve

flushing valve, operated modulely or by automatic means

150mm minimum

pipe interrupter with atmospheric vent for backflow protection

urinal bowl

urinal bowl

flush pipe

150mm minimum

Pressure flushing valves, operating modulely or by automatic means e.g. remote sensor – maximum delivery 1.5 litres per operation

Pressure flushing valves may be supplied from the supply pipe or from distributing pipes

Connections to urinals using pressure flushing valves may be made from a supply pipe or a distributing pipe, but when installing them **a permanently vented pipe interrupter must be fitted at the outlet of the flushing valve and at least 150mm above the discharge outlet into the urinal bowl.**

Individual urinal flushing (using flushing valve)

This new method of flushing an individual bowl after use with a flushing valve (no cistern required) has the flowing advantages:

- Maximum water saving – individually flushed
- Minimum smell – flushes automatically **immediately** after use
- No bulky cistern or large pipes – pipes are 15mm and normally concealed
- No overflow required
- Flush volume and flow rate easily altered at any time
- No small pet cocks or small orifices to block
- No planned maintenance
- Reliable (individual isolation if maintenance is required)
- Hygienic and visually impressive

Should we use water economy controls for urinals?

Yes! The requirement of Paragraph 25(1)(j) is that any urinal supplied either modulely or electronically from a flushing cistern, must have a time switch (and a lockable isolating valve) fitted to its incoming supply, or some other equally effective means of regulating the periods during which the cistern may fill.

Many buildings of course are only occupied at certain times of the day, and often not every day of the week.

Satisfactory regulation of water flow to urinals can be achieved in a number of ways, which may include, time control or hydraulic flow control devices which will ensure that the flushing cistern will only fill during periods when the urinals are being used.

These devices are illustrated in the following diagrams.

Flushing urinals from a cistern

This conventional arrangement consists of a cistern mounted above a single or group of urinals. Water gradually fills the cistern either at a continuous slow rate (through a 'pet cock') or modulated (turned on or off) at a higher flow rate. The cistern flushes once the water level reaches the self-siphon level.

Automatic Urinal Filling Rate, Schedule 2, Paragraph 25 states the legal requirements for WCs and urinals. This is repeated and reinforced by the two guidance clauses reproduced below.

Clause G25.12 states:

An automatically operated flushing cistern serving urinals should be filled with water at a rate not exceeding:

(a) *10 litres per hour per urinal bowl for a cistern serving a single urinal: or*

(b) *7.5 litres per hour per urinal bowl or position or, as the case may be, for each 700mm width of urinal slab for a cistern serving two or more urinals.*

Clause G25.10 states:

Unless a urinal cistern is manually operated, or fills and flushers by a device operated by an electronic sensor, pressure pad or no less suitable device which ensures that the urinal is only flushed after it is used, the inlet to the flushing cistern is to be controlled by a time switch opening an inlet valve or some other equally effective automatic device which regulates the periods during which the cistern may fill.

There are three acceptable methods for preventing flushing when urinals are not being used:

- Time switch and pet cock-this is a crude arrangement with many inherent limitations
- Hydraulic impulse device (original Cistermiser) – these devices rely on a man washing his hands after using the urinal. Providing the wash basin tap is taken off the same circuit as the device it should initiate a mechanical timer. If the user does not wash his hands or uses the hot tap the device will not work. Reliability is impaired with high calcium carbonate in the water. Electronic sensors are much better
- Electronic movement detectors – this method is technically superior to others. They can be battery operated (often providing more than two years life) or mains operated. Devices are available for controlling conventional flushing cisterns or flushing valves (no cistern required).

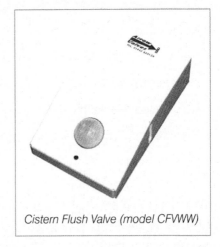

Cistern Flush Valve (model CFVWW)

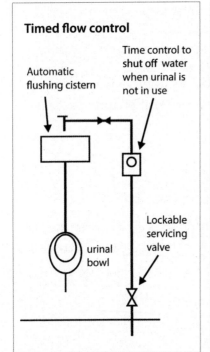

Timed flow control

Automatic flushing cistern

Time control to shut off water when urinal is not in use

urinal bowl

Lockable servicing valve

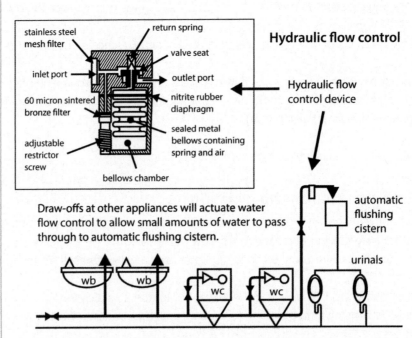

Hydraulic flow control

stainless steel mesh filter

return spring

valve seat

inlet port

outlet port

60 micron sintered bronze filter

nitrite rubber diaphragm

adjustable restrictor screw

sealed metal bellows containing spring and air

bellows chamber

Hydraulic flow control device

Draw-offs at other appliances will actuate water flow control to allow small amounts of water to pass through to automatic flushing cistern.

wb   wb   WC   WC

automatic flushing cistern

urinals

How can we work out the size of cistern we need and the interval between flushing to meet the requirements of the regulations?

Previous discussions have looked at the requirements for flushing urinals from cisterns in terms of volume per individual bowl, stall or slab, but it doesn't tell us how to set the flow rate to achieve the requirements. So! Let's look at time intervals between flushing using standard sized flushing cisterns. i.e. cisterns of *4.5, 9, 13.5, and 18 litre* capacity.

First a calculation to see how to work out the time interval between flushing, then we provide for your use, a table of time intervals based on the calculations we have shown.

Calculation of flushing intervals

For example using a 13.5 litre automatic cistern to serve five urinal bowls:

$$\text{Time interval (minutes)} = \frac{\text{Cistern capacity (litres) X time in one hour (60 min.)}}{\text{maximum flush requirements X No. of urinal positions served}}$$

$$\text{Time interval} = \frac{\text{Cistern capacity (13.5 litres) X time in one hour (60 min.)}}{\text{maximum flush (7.5 litres per bowl) X No. of bowls (5)}}$$

$$= \frac{13.5 \text{ X } 60}{7.5 \text{ X } 5} = \textbf{21.6 minutes}$$

So! A flush interval of 21.6 minute s will achieve the requirement of 7.5 litres per hour, per bowl, for five urinal bowls supplied from a 13.5 litre capacity automatic flushing cistern.

Volumes and flushing intervals for urinals

Number of bowls, stalls or per 700mm of slab	Volume of automatic flushing cistern				Maximum fill rate in litres per hour
	4.5 litres	9 litres	13.5 litres	18 litres	
	Shortest period between flushes				
1	27	54	81	108	10
2	18	37	54	72	15
3	12	24	36	48	22.5
4	9	18	27	36	30
5	7.2	14.4	21.6	28.8	37.5
6	6	12	18	24	45

Finally, a summary of permitted flushing volumes for urinals is shown in the following table.

Maximum permitted flushing volumes for urinals

Appliance	Maximum permitted volume
Single bowl or stall supplied from an automatic flushing cistern	10 litres per hour per bowl
More than one appliance supplied from an automatic flushing cistern	7.5 litres per hour per bowl, stall or per 700mm width of slab
Single or multiple bowl or stall installations operated modulely or automatically	1.5 litres per flush as required

Self-assessment questions

1. State the main purpose of Paragraph 25 of Schedule 2.

   _____

2. There are a number of new provisions in Paragraph 25 of Schedule 2.
   Give THREE of these new provisions:

   1. _____

   2. _____

   3. _____

3. Complete the diagram below to indicate the backflow protection that should be provided to a pressure flushing valve for use with a WC cistern.

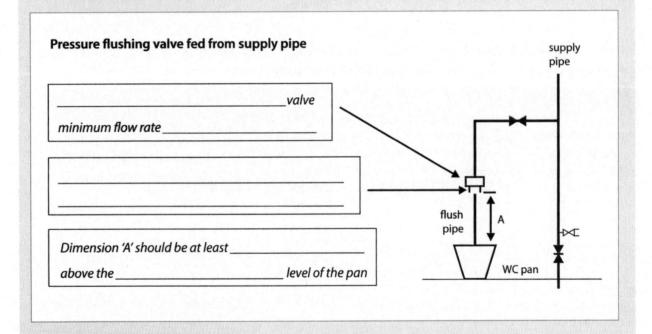

Pressure flushing valve fed from supply pipe

_____ valve

minimum flow rate _____

_____

_____

Dimension 'A' should be at least _____

above the _____ level of the pan

supply pipe

flush pipe   A

WC pan

4. State the maximum capacities required by the Regulations for WC cisterns.

   a) single flush cistern _____

   b) a dual flush cistern   – full flush _____

                             – reduced flush _____

5. State what markings are required to be shown on dual flushing cisterns.

_____

6. State the requirement for the normal termination of a warning pipe from a WC cistern.

_____

_____

7. It is permitted to use common warning pipes for multiple installations of WC cisterns providing certain rules are adhered to.

State the rules in the spaces below.

_____

_____

_____

8. Can the overflow pipe be discharged into the WC pan?

_____

_____

_____

9. Assume you are called in to replace an old damaged 9 litre WC cistern, but the pan is in good condition. Under Water Regulations you have two options open to you. Give the TWO options:

1. _____

_____

2. _____

_____

10. The following diagrams illustrate three separate urinal flushing arrangements using flushing cisterns. In the spaces provided, show the maximum permitted flush volume in each case.

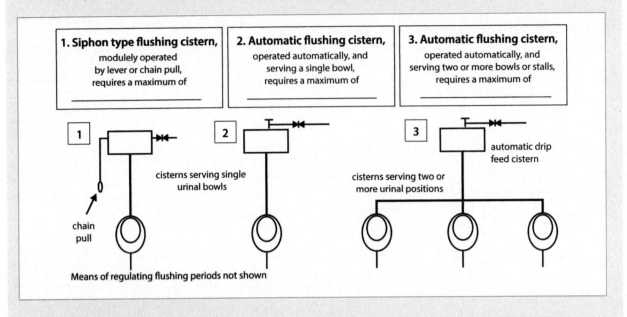

1. Siphon type flushing cistern, modulely operated by lever or chain pull, requires a maximum of

_____

2. Automatic flushing cistern, operated automatically, and serving a single bowl, requires a maximum of

_____

3. Automatic flushing cistern, operated automatically, and serving two or more bowls or stalls, requires a maximum of

_____

cisterns serving single urinal bowls

cisterns serving two or more urinal positions

automatic drip feed cistern

chain pull

Means of regulating flushing periods not shown

11. Give THREE methods permitted by Regulations for controlling the flow to urinals that will ensure they only use water during periods of use.

1. _____

2. _____

3. _____

12. The diagrams below illustrate urinal flushing arrangements using pressure flushing valves. Complete the labelling to show minimum permitted flush volumes and suitable backflow protection requirements.

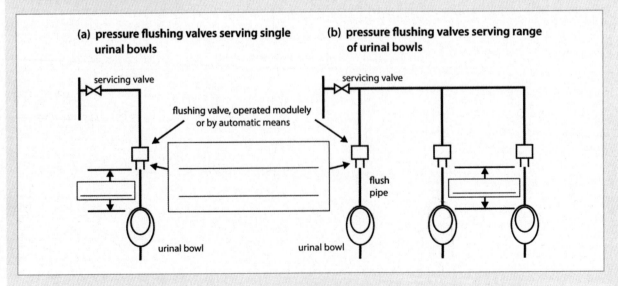

(a) **pressure flushing valves serving single urinal bowls**

(b) **pressure flushing valves serving range of urinal bowls**

servicing valve

servicing valve

flushing valve, operated modulely or by automatic means

flush pipe

urinal bowl

urinal bowl

Check your answers on pages 21, 22 and 23.

Summary of main points

Paragraph 25 of Schedule 2 is concerned primarily with the **conservation of water**.

New provisions **reduce flushing volume for washdown WCs** and the **dual flush cistern is back** in fashion. Non siphonic mechanisms are now permitted in WC cisterns, and **pressure flushing valves are permitted for both WCs and urinals but not in a house.**

WCs cisterns may be flushed by **single** or **dual flush mechanisms**, using **siphonic action** or by **non-siphonic action** (drop valve mechanism).

WCs may be flushed using a pressure flushing valve, but NOT in a house. Flushing valves need flow rate of at least **1.2 litres per second**.

Flushing volumes to WC pans (after 1 January 2001) must not exceed:

- **six litres maximum** for single flush cisterns or dual flush cisterns

- **four litres maximum** for the **reduced flush** in a dual flushing cistern

A dual flush cistern should have its **operating procedure/mechanism clearly and permanently marked on or near the cistern.**

When replacing a cistern or pan, it must be of the same design as the original, otherwise it is unlikely that they will operate satisfactorily.

Alternatively, both cistern and pan should be changed for a new suite.

When **external warning pipes**, are used on WC flushing cisterns they **must discharge where they can readily be seen.**

Individual warning pipes from WC cisterns **may** discharge into a **common warning pipe serving WCs only** that are **all at the same level**, and the **individual warning pipes should discharge into a tundish** which is **readily visible**. A tundish is used so that it can be established which cistern requires attention.

If the discharge is to a pan, **it must discharge into the air** at least **150mm above the rim** of the pan.

Overflow pipes, an internal overflow discharging into the WC pan can be used, provided that a **Type AG Air Gap** is provided between the lowest level of the outlet of the float valve and the discharge from the overflow pipe.

Internal overflow pipes, for compact inlet valves to BS 1212: Part 4 the provision of a gauze strainer incorporated in, or fitted upstream of, the float-operated valve to trap debris which might cause premature failure of the valve is required.

Urinals may be flushed by a **cistern** that is **modulely** or **automatically** operated, or by a **flushing valve** operated **modulely** or **automatically**.

Flushing cisterns, operated automatically, require:

- **10 litres per hour for a single urinal bowl,** or stall or slab position. (700mm);

- **7.5 litres per hour** (per urinal position) **for a cistern serving two or more urinal bowls,** stalls or slab positions

Modulely operated cisterns serving a single bowl or stall require:

- **1.5 litres per operation.**

Pressure flushing valves operated either modulely or automatically require:

- **not more than 1.5 litres** each time the valve is operated

Urinals should flush only during periods of use! Devices to regulate water flow to urinals may include, time control, photo-electric cells, pressure pad control, hydraulic flow control or simple methods of module operation such as the chain pull and siphon.

Pressure flushing valves used for flushing WCs or urinals **may be connected to a supply pipe or a distributing pipe** and for backflow protection, they are required to be **fitted with a permanently vented pipe interrupter** at the outlet of the pressure flushing valve. As you will recall from Module 8, WC pans and urinals are considered to be a fluid category 5 backflow risk.

The pipe interrupter should be located:

- at the outlet of the flushing valve;

- **not less than 300mm above the spillover level of the pan, or**

- **not less than 150mm above the discharge outlet into the urinal bowl; and**

- **there should be no tap, valve or restriction to the flow at its outlet.**

Answers to self-assessment questions

1. State the main purpose of Paragraph 25 of Schedule 2.

   Conservation of water

2. There are a number of new provisions in Paragraph 25 of Schedule 2 Give THREE of these new provisions.

   1. *reduced flushing in WCs*
   2. *reintroduced dual-flush cisterns*
   3. *non-siphonic flushing cisterns permitted for WC flushing valves permitted for WCs and urinals*

3. Complete the diagram below to indicate the backflow protection that should be provided to a pressure flushing valve for use with a WC cistern.

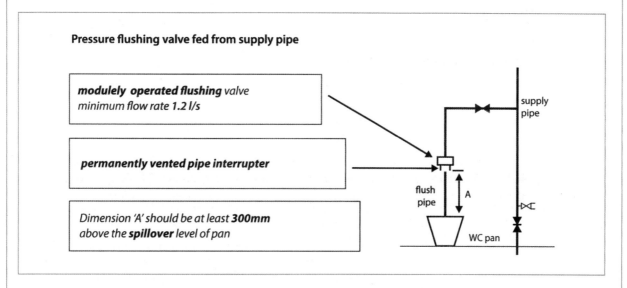

Pressure flushing valve fed from supply pipe

modulely operated flushing valve minimum flow rate 1.2 l/s

permanently vented pipe interrupter

Dimension 'A' should be at least 300mm above the spillover level of pan

4. State the maximum capacities required by the Regulations for WC cisterns.

   a) *single flush cistern*   *6 litres*

   b) *a dual flush cistern*   *– full flush*   *6 litres*

                             – reduced flush   *4 litres*

5. State what markings are required to be shown on dual flushing cisterns.

   Instructions clearly and permanently marked on or near cistern

6. State the requirement for the normal termination of a warning pipe from a WC cistern.

   It must be located in a position where any discharge can be readily seen

7. It is permitted to use common warning pipes for multiple installations of WC cisterns providing certain rules are adhered to.

    State the rules in the spaces below.

    WCs all at one level

    Discharge should be via a tundish, and
    the discharge to each tundish must be readily visible to help establish which cistern requires attention

8. Can the overflow pipe be discharged into the WC pan?

    Yes. *Provided that a Type AG Air Gap is provided between the lowest level of the outlet of the float valve and the discharge from the overflow pipe.*

9. Assume you are called in to replace an old, damaged 9 litre WC cistern, but the pan is in good condition. Under Water Regulations you have two options open to you.

    Give the TWO options:

    1. *It must be of the same design as the original one, or*

    2. *both cistern and pan should be replaced*

10. The following diagrams illustrate three separate urinal flushing arrangements using flushing cisterns. In the spaces provided, show the maximum permitted flush volume in each case.

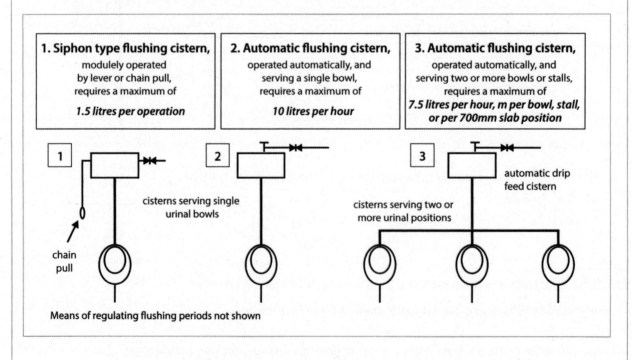

1. Siphon type flushing cistern,	2. Automatic flushing cistern,	3. Automatic flushing cistern,
modulely operated by lever or chain pull, requires a maximum of	operated automatically, and serving a single bowl, requires a maximum of	operated automatically, and serving two or more bowls or stalls, requires a maximum of
1.5 litres per operation	**10 litres per hour**	**7.5 litres per hour, m per bowl, stall, or per 700mm slab position**

chain pull

cisterns serving single urinal bowls

automatic drip feed cistern

cisterns serving two or more urinal positions

Means of regulating flushing periods not shown

11. Give THREE methods permitted by Regulations for controlling the flow to urinals that will ensure they only use water during periods of use.

1. *pressure flushing valve*

2. *time switch*

3. *hydraulic flow control*

12. The diagrams below illustrate urinal flushing arrangements using pressure flushing valves. Complete the labelling to show minimum permitted flush volumes and suitable backflow protection requirements.

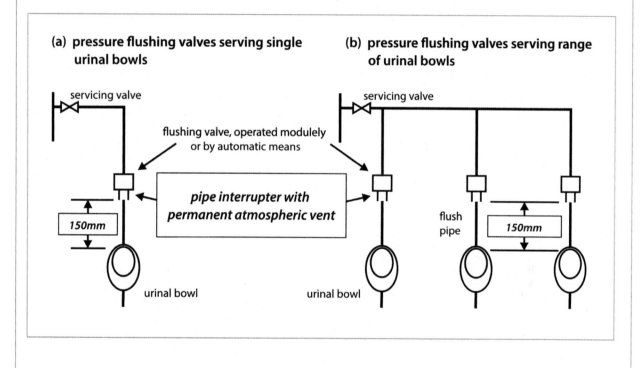

(a) pressure flushing valves serving single urinal bowls

servicing valve

flushing valve, operated modulely or by automatic means

pipe interrupter with permanent atmospheric vent

150mm

urinal bowl

(b) pressure flushing valves serving range of urinal bowls

servicing valve

flush pipe

150mm

urinal bowl

What to do next

So much for WCs and urinals!

Hope it didn't give you a hot flush?

Now go on to **Module 12** Sanitary Appliances and Water for Outside Use

Water Industry Act 1991:

Water Supply (Water Fittings) Regulations 1999

An Open Learning Course

Module 12

Sanitary appliances and water for outside use

Introduction

This module looks at Paragraphs 26 to 31 of Schedule 2 to the Regulations which deal with the prevention of contamination and waste of water in a variety of topics relating to the supply of water to sanitary appliances and water for outside use.

Topics included are:

- the supply of drinking water for domestic purposes
- waste outlet and plugs for baths, sinks showers and taps, and
- quantities of water used in washing machines/driers and dishwashers
- water to animal drinking vessels
- ponds, fountains and pools

The requirements are generally quite clear and straightforward. There is little that is new in Paragraphs 26 to 31 as much of what is contained in them has been required by water byelaws for many years.

One new provision however, is seen in Paragraph 29 which introduces further economy measures by requiring washing machines and dishwashers to use less water than was previously permitted.

What is the requirement?

Schedule 2: Paragraphs 26, 27 and 28

Baths, sinks, showers and taps

26. *All premises supplied with water for domestic purposes shall have at least one tap conveniently situated for the drawing of drinking water.*

27. *A drinking water supply shall be supplied with water from:*

  (a) *a supply pipe;*

  (b) *a pump delivery pipe drawing water from a supply pipe; or*

  (c) *a distributing pipe drawing water exclusively from a storage cistern supplying wholesome water*

28.-(1) *Subject to Paragraph (2), every bath, wash basin, sink or similar appliance shall be provided with a watertight and readily accessible plug or other device capable of closing the waste outlet.*

  (2) *This requirement does not apply to:*

  (a) *an appliance where the only taps provided are spray taps;*

  (b) *a washing trough or wash basin whose waste outlet is incapable of accepting a plug and to which water is delivered at a rate not exceeding 0.06 litres per second exclusively from a fitting designed or adapted for that purpose;*

  (c) *a wash basin or washing trough fitted with self closing taps;*

  (d) *a shower bath or shower tray;*

  (e) *a drinking water fountain or similar facility; or*

  (f) *an appliance which is used in medical, dental or veterinary premises and is designed or adapted for use with an unplugged outlet.*

Schedule 2: Paragraph 29

Washing machines, dishwashers and other appliances

29.-(1) *Subject to Paragraph (2), clothes washing machines, clothes washer-driers and dishwashers shall be economical in the use of water.*

  (2) *The requirements of this paragraph shall be deemed to be satisfied in the case of machines having a water consumption per cycle of not greater than the following:*

  (a) *For domestic horizontal axis washing machines, 27 litres per kilogram of wash load for a standard 60°C cotton cycle;*

  (b) *For domestic washer-driers, 48 litres per kilogram of wash load for a standard 60°C cotton cycle;*

  (c) *For domestic dishwashers, 4.5 litres per place setting*

Paragraphs 30 and 31 are continued on the next page

> **Schedule 2: Paragraph 30 and 31**
>
> **Water for outside use**
>
> 30.   *Every pipe which conveys water to a drinking vessel for animals or poultry shall be fitted with:*
>
>   *(a) a float operated valve or some other no less effective device to control the inflow of water, which is:*
>
>   *(i) protected from damage and contamination; and*
>
>   *(ii) prevents contamination of the water supply; and*
>
>   *(b) a stopvalve or servicing valve as appropriate*
>
> 31.   *Every pond, fountain or pool shall have an impervious lining or membrane to prevent the leakage or seepage of water.*

What is important about drinking water supplies to domestic premises?

This is dealt with in Paragraphs 26 and 27.

'All premises supplied with water for domestic purposes shall have at least one tap conveniently situated for the drawing of drinking water'.

To clarify which premises are required to have a drinking water tap fitted, it is perhaps worth looking at definition of 'domestic purposes'.

The Water Industry Act 1991 (paragraph 218) says that in general:

- 'water for domestic purposes means water used for:
  - **drinking**
  - **washing**
  - **cooking**
  - **central heating, and**
  - **sanitary purposes'**

Additionally, domestic purposes includes the **washing of vehicles and watering of gardens** but not with a hosepipe.

A drinking water supply shall be supplied with water from:

(a) a supply pipe;

(b) a pump delivery pipe drawing water from a supply pipe; or

(c) a distributing pipe drawing water exclusively from a storage cistern supplying wholesome water.

These pipes are illustrated in the following diagrams:

Drinking water supply to single dwelling

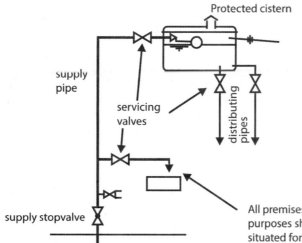

Protected cistern

supply pipe

servicing valves

distributing pipes

supply stopvalve

A drinking water supply shall be supplied with water from: a supply pipe; a pump delivery pipe drawing water from a supply pipe; or a distributing pipe drawing water exclusively from a storage cistern supplying wholesome water

A distributing pipe may be used to supply drinking water providing the storage cistern is protected from contamination

All premises supplied with water for domestic purposes shall have at least one tap conveniently situated for the drawing of drinking water. In dwellings this is usually at the kitchen sink.

Boosted supply drawing cold water draw-off points from a supply pipe and from a pump delivery pipe and including a 'protected cistern'

A drinking water supply shall be supplied with water from: a supply pipe; a pump delivery pipe drawing water from a supply pipe; or a distributing pipe drawing water exclusively from a storage cistern supplying wholesome water

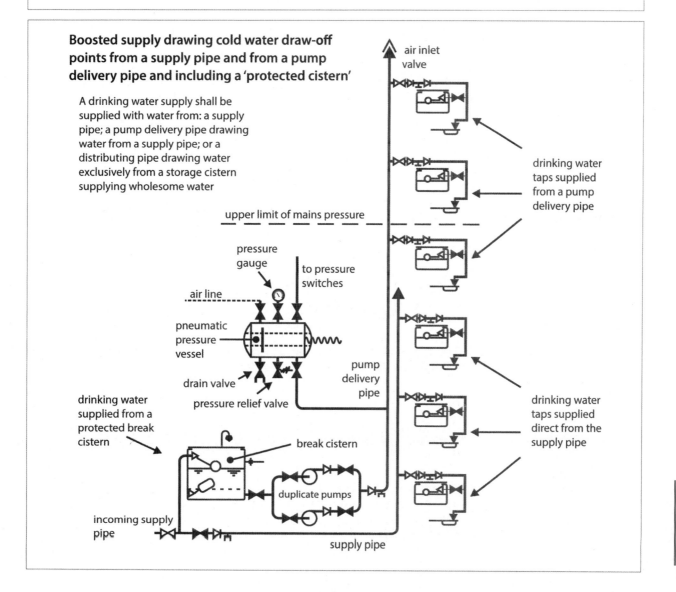

air inlet valve

drinking water taps supplied from a pump delivery pipe

upper limit of mains pressure

pressure gauge

to pressure switches

air line

pneumatic pressure vessel

drain valve

pressure relief valve

pump delivery pipe

drinking water supplied from a protected break cistern

break cistern

incoming supply pipe

duplicate pumps

supply pipe

drinking water taps supplied direct from the supply pipe

Boosted system showing drinking water taps supplied from a pumped delivery pipe and a distributing pipe from a 'protected cistern'

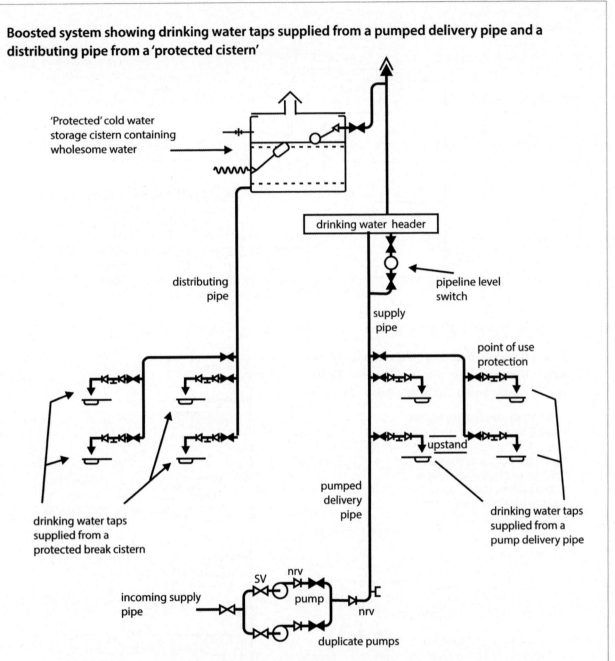

'Protected' cold water storage cistern containing wholesome water

drinking water header

distributing pipe

pipeline level switch

supply pipe

point of use protection

drinking water taps supplied from a protected break cistern

pumped delivery pipe

upstand

drinking water taps supplied from a pump delivery pipe

incoming supply pipe

SV   nrv

pump

nrv

duplicate pumps

A drinking water supply shall be supplied with water from: a supply pipe; a pump delivery pipe drawing water from a supply pipe; or a distributing pipe drawing water exclusively from a storage cistern supplying wholesome water

What are the requirements of Paragraph 28 for sanitary appliances?

There is one simple requirement in Paragraph 28(1) which states that every bath, wash basin, sink, and other similar appliance **must have a plug fitted to its waste outlet**. The plug must be **watertight** and **accessible**. The usual rubber/plastic push-in plug attached by a chain is quite adequate, as is the lever operated plug working through combination tap arrangement. However, if a lever operated plug is used there must be no connection made hydraulically between the water supply and the waste fitting.

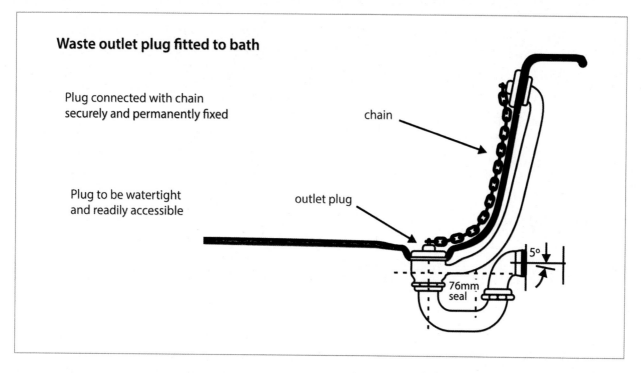

Waste outlet plug fitted to bath

Plug connected with chain securely and permanently fixed

chain

Plug to be watertight and readily accessible

outlet plug

76mm seal

5°

Are there any exceptions?

Yes! Paragraph 28(2) permits a number of exceptions, and these are all appliances that are fitted with low flow taps or inlet valves.

An outlet plug is NOT needed for:

(a) an appliance where the only taps provided are spray taps; e.g. wash basins

(b) a washing trough or wash basin to which a plug a cannot be fitted and to which water is delivered at a rate not exceeding 0.06 litres per second

(c) a wash basin or washing trough fitted with self closing taps

(d) a shower bath or shower tray

(e) a drinking water fountain or similar facility; or

(f) an appliance which is used in medical, dental or veterinary premises and is designed or adapted for use with an unplugged outlet.

Are there any new rules relating to clothes washing machines and dish washing machines?

Paragraph 29 is included as a conservation measure. It says clothes washing machines, clothes washer-driers and dishwashers shall be economical in the use of water, and it limits the quantity of water permitted for use in washing machines and dishwashers.

The following table sets out the requirements of Paragraph 29.

Maximum water consumption for washing machines and dishwashers	
Domestic horizontal axis *washing machines*	*27 litres per kilogram of wash load for a standard 60°C cotton cycle*
Domestic *washer-driers*	*48 litres per kilogram of wash load for a standard 60°C cotton cycle*
Domestic *dishwashers*	*4.5 litres per place setting*

Additional information

Since the introduction of more efficient appliances the water consumption levels have greatly reduced.

The average washing machine now uses around 11.5 litres of water for every kilogram of cottons it washes on the standard 40°C cotton program (depending on model).

The average washer-dryer uses a similar amount of water to wash as a washing machine (11.5 litres of water for every kilo of clothes washed), but you may not know that it also uses an average of around six litres of water per kilogram of washing for cooling during the dryer cycle. Combining the two figures this is still considerably less than the 48 litres previously mentioned. But the amount of water used by different washer-dryers varies considerably.

Government buying standard for dishwashers September 2015

Mandatory level – water consumption for a standard sized dishwasher (for up to 14 place settings) volume per cycle must not exceed 10 litres (range 0.67 to 1.25 litres/place setting).

Best practice – level water consumption in a standard sized dishwasher (for up to 14 place settings) volume per cycle must not exceed 8 litres

Again this is considerably less than the 4.5 litres previously mentioned.

We have already dealt with methods of contamination prevention for these machines in Module 4 backflow prevention.

Now to our final topic in the learning package, 'water for outside use'.

What further requirements are there for water for outside use?

Paragraph 30 looks at animal drinking troughs and bowls and gives further requirements for the provision of stopvalves and servicing valves and for the prevention of contamination.

Much of this has already been dealt with in Modules 5 and 8, but Paragraph 30 is quite specific about animal drinking vessels. We should look again at them now.

The following important points are made:

(a) **Pipes supplying water to drinking vessels for animals must be fitted with a float operated valve or some other equally effective device** for controlling the inflow of water to the vessel, and:

   ii)   **the flow control device must be protected from damage and contamination;**

   iii) **there must be means of preventing contamination of the water supply.**

(b) **Pipes supplying water to drinking vessels for animals must be fitted with a stopvalve or servicing valve, as appropriate.**

These points are illustrated in the following diagrams:

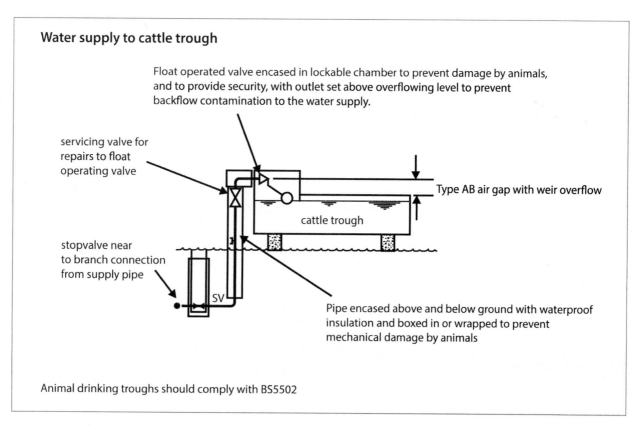

Water supply to cattle trough

Float operated valve encased in lockable chamber to prevent damage by animals, and to provide security, with outlet set above overflowing level to prevent backflow contamination to the water supply.

servicing valve for repairs to float operating valve

Type AB air gap with weir overflow

cattle trough

stopvalve near to branch connection from supply pipe

SV

Pipe encased above and below ground with waterproof insulation and boxed in or wrapped to prevent mechanical damage by animals

Animal drinking troughs should comply with BS5502

Individual drinking bowls will need to treated differently depending on whether their outlet nozzle is arranged with a Type AA air gap or if its outlet is submerged it will need to be supplied from storage using dedicated distributing pipes that only supply that type of appliance.

Animal drinking bowls

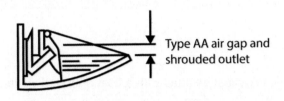

Type AA air gap and shrouded outlet

This type with spring return valve or float valve operated when depressed by an animal's mouth may be connected directly from a supply pipe or distributing pipe providing the Type AA air gap is maintained and animal's mouth cannot come into contact with the outlet nozzle.

Nozzle outlet below spillover level

Where outlet nozzle is below spillover level or is likely to be contaminated by animal's mouth the bowl must be supplied from a dedicated cistern that only supplies similar appliances.

Finally! Ponds, fountains or pools.

What is important about these ponds, fountains or pools?

Very simple this one. **They must have an impervious lining and must not leak!**

This applies to ALL ponds pools or fountains, not just large ones and irrespective of their use.

Large pools, because they are more costly to build and maintain, often more care is taken to ensure they are watertight and will remain so. Smaller ones on the other hand are often built quickly with less care, particularly those built on a DIY basis.

Pools or ponds used for industrial purposes, fire supplies and the like are just as important as the swimming pool or garden feature in a home. They will all waste water if they leak!

Two final points to remember:

1. Ponds pools or fountains should not be connected to a supply pipe or distributing pipe. We dealt with that in Module 8 'backflow'.

2. Don't forget that the installation of any pool of more than 10,000 litres capacity which is replenished by automatic means is required under Regulation 4(5) to be notified to the water undertaker before the installation is commenced.

Self-assessment questions

1. Paragraph 26 requires premises supplied with water for domestic purposes to have one tap conveniently situated for the drawing of drinking water.

   Of the following, which are considered to be 'water used for domestic purposes':

   Water for washing   Yes ☐   No ☐

   Water for central heating   Yes ☐   No ☐

   Water used for washing a car by hosepipe at a house   Yes ☐   No ☐

   Water used for car washing in business premises   Yes ☐   No ☐

2. Drinking water is permitted to be supplied from three pipe systems only.
   Name the THREE pipe systems.

   a) _____

   b) _____

   c) _____

3. Paragraph 28 gives one simple rule for sanitary appliances. They must each be fitted with a waste outlet plug. Some appliances however, are excepted from this rule. Indicate, by *Yes* or *No*, to which of the following appliances Paragraph 28 applies.

   bath ☐   *drinking water fountain* ☐

   wash basin ☐   *basin with spray taps* ☐

   sink ☐   *hand washing trough in industrial premises* ☐

4. Paragraph 29 sets out maximum water usage in washing machines and dishwashers. Complete the following table of maximum water consumptions:

Maximum water consumption for washing machines and dishwashers	
Domestic horizontal axis washing machines	_____ of wash load for a standard 60°C cotton cycle
Domestic washer-driers	_____ of wash load for a standard 60°C cotton cycle
Domestic dishwashers	_____ per place setting

5. Paragraph 30 deals with animal drinking bowls and troughs, and requires measures to be taken to control the inflow of water, to protect against mechanical damage, and guard against the possibility of contamination.

Annotate the following diagrams, to show how these requirements can be achieved.

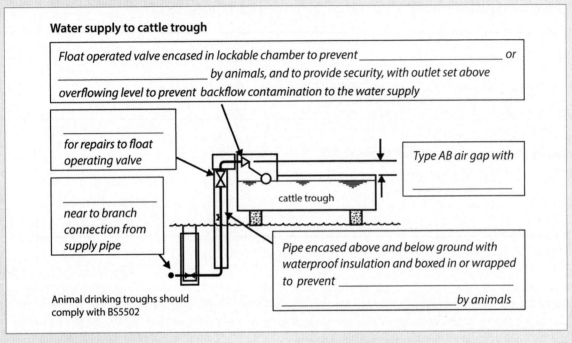

Water supply to cattle trough

Float operated valve encased in lockable chamber to prevent _____ or _____ by animals, and to provide security, with outlet set above overflowing level to prevent backflow contamination to the water supply

_____ for repairs to float operating valve

_____ near to branch connection from supply pipe

Animal drinking troughs should comply with BS5502

cattle trough

Type AB air gap with _____

Pipe encased above and below ground with waterproof insulation and boxed in or wrapped to prevent _____ _____by animals

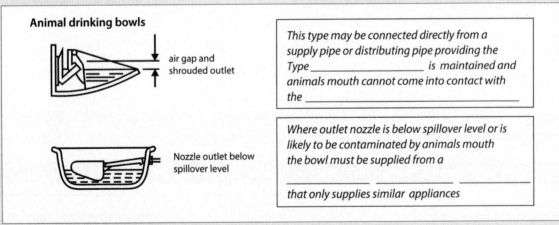

Animal drinking bowls

air gap and shrouded outlet

Nozzle outlet below spillover level

This type may be connected directly from a supply pipe or distributing pipe providing the Type _____ is maintained and animals mouth cannot come into contact with the _____

Where outlet nozzle is below spillover level or is likely to be contaminated by animals mouth the bowl must be supplied from a

_____ _____ _____

that only supplies similar appliances

6. Complete the following statements that follow guidance given in Clauses 5.3.5./6./7.

Ponds, fountains and pools (including fish ponds and swimming pools) must:

a) be _____ with some sort of _____ lining

b) not be _____ _____ to any supply pipe or distributing pipe, and

c) if more than 10,000 litres capacity, be _____ to the water undertaker before installation work is started.

Check your answers on pages 14 and 15.

Summary of main points

'All premises supplied with water for domestic purposes shall have at least one tap conveniently situated for the drawing of drinking water'.

In general 'water for domestic purposes' means water used for drinking, washing, cooking, central heating, and sanitary purposes.

Additionally 'water for domestic purposes' includes the washing of vehicles and watering of gardens but not with a hosepipe.

A drinking water supply shall be supplied with water from:

(a) a supply pipe;

(b) a pump delivery pipe drawing water from a supply pipe; or

(c) a distributing pipe drawing water exclusively from a storage cistern supplying wholesome water

Every bath, wash basin, sink, and other similar appliance **must have a plug fitted to its waste outlet**. The plug must be **watertight** and **accessible**, but there are some **exceptions. An outlet plug is NOT needed for:**

(a) an appliance with spray taps

(b) a washing trough or wash basin to which a plug a cannot be fitted and water is delivered at a rate not exceeding 0.06 litres per second

(c) a wash basin or washing trough fitted with self closing taps

(d) a shower bath or shower tray

(e) a drinking water fountain or similar; or

(f) an appliance which is used in medical, dental or veterinary premises and is designed or adapted for use with an unplugged outlet.

Maximum water consumption for washing machines and dishwashers	
Domestic horizontal axis *washing machines*	*27 litres per kilogram of wash load for a standard 60°C cotton cycle*
Domestic *washer-driers*	*48 litres per kilogram of wash load for a standard 60°C cotton cycle*
Domestic dishwashers	*4.5 litres per place setting*

- Animal and poultry drinking troughs **must be fitted with float operated** valve or other effective shut-off device, and **measures taken to prevent damage and to provide protection from contamination.**

- Ponds, fountains and pools, including fish ponds and swimming pools must be **watertight** with some sort of **impervious lining.** They must **not be directly connected to any supply pipe or distributing pipe.**

Answers to self-assessment questions

1. Paragraph 26 requires premises supplied with water for domestic purposes to have one tap conveniently situated for the drawing of drinking water.

   Of the following, which are considered to be 'water used for domestic purposes':

Water for washing	Yes ✔	No
Water for central heating	Yes ✔	No
Water used for washing a car by hosepipe at a house	Yes	No ✔
Water used for car washing in business premises	Yes	No ✔

2. Drinking water is permitted to be supplied from three pipe systems only. Name the THREE pipes systems are:

   a) **a supply pipe;**

   b) **a pump delivery pipe drawing water from a supply pipe; or**

   c) **a distributing pipe drawing water exclusively from a storage cistern supplying wholesome water**

3. Paragraph 28 gives one simple rule for sanitary appliances. They must each be fitted with a waste outlet plug. Some appliances however, are excepted from this rule.

   Indicate, by *Yes* or *No*, to which of the following appliances Paragraph 28 applies.

bath	Yes	drinking water fountain	No
wash basin	Yes	basin with spray taps	No
sink	Yes	hand washing trough in industrial premises	No

4. Maximum water usage in washing machines and dishwashers is set out in Paragraph 29. Complete the following table of maximum water consumptions:

Maximum water consumption for washing machines and dishwashers	
Domestic horizontal axis *washing machines*	*27 litres per kilogram of wash load for a standard 60°C cotton cycle*
Domestic *washer-driers*	*48 litres per kilogram of wash load for a standard 60°C cotton cycle*
Domestic dishwashers	*4.5 litres per place setting*

5. Paragraph 30 deals with animal drinking bowls and troughs, and requires measures to be taken to control the inflow of water, to protect against mechanical damage, and guard against the possibility of contamination.

Annotate the following diagrams, to show how these requirements can be achieved.

Water supply to cattle trough

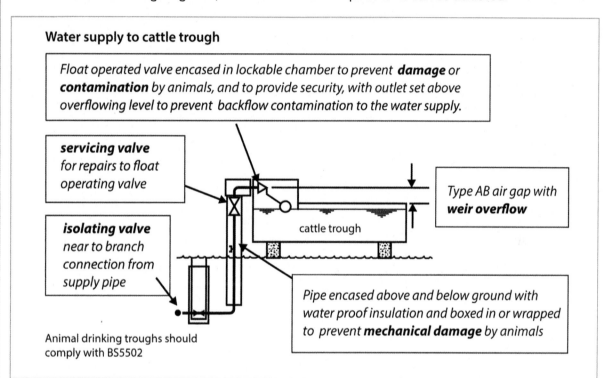

*Float operated valve encased in lockable chamber to prevent **damage** or **contamination** by animals, and to provide security, with outlet set above overflowing level to prevent backflow contamination to the water supply.*

servicing valve *for repairs to float operating valve*

isolating valve *near to branch connection from supply pipe*

*Type AB air gap with **weir overflow***

cattle trough

*Pipe encased above and below ground with water proof insulation and boxed in or wrapped to prevent **mechanical damage** by animals*

Animal drinking troughs should comply with BS5502

Animal drinking bowls

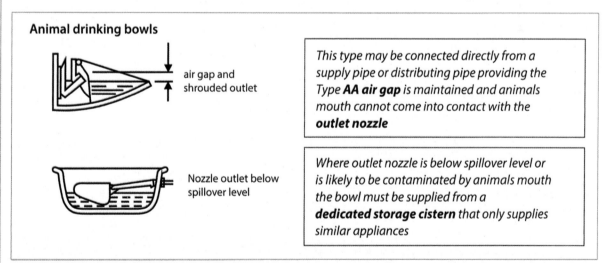

air gap and shrouded outlet

*This type may be connected directly from a supply pipe or distributing pipe providing the Type **AA air gap** is maintained and animals mouth cannot come into contact with the **outlet nozzle***

Nozzle outlet below spillover level

*Where outlet nozzle is below spillover level or is likely to be contaminated by animals mouth the bowl must be supplied from a **dedicated storage cistern** that only supplies similar appliances*

6. Complete the following statements that follow guidance given in Clauses 5.3.5./6./7.

Ponds, fountains and pools (including fish ponds and swimming pools) must:

a) *be **watertight** with some sort of **impervious** lining,*

b) *not be **directly connected** to any supply pipe or distributing pipe, and*

c) *if of more than 10,000 litres capacity, be **notified** to the water undertaker before installation work is started.*

What to do next

Well done!

You have finished your course of study.

You should now consider Assessment

Have a go at the **Multiple Choice Question Papers** and the **Practical Task Assessment Video** at your nearest BPEC Training and Assessment Centre

Phone 0845 307 6105 for further details

Water Industry Act 1991:

Water Supply (Water Fittings) Regulations 1999

An Open Learning Course

Glossary of Terms

To support the self learning course

Introduction

Throughout your study of the Water Regulations you will come across a number of terms, definitions and interpretation, some of which may need to be explained. It is the purpose of this document to do just that. Much of this duplicates what is written in various parts of your manuals, but as you go through your studies and come across a term, it will often be easier to check its meaning by looking it up in this Glossary rather than wade through the manuals to find it.

The Water Supply (Water Fittings) Regulations 1999, is a legal document and must be interpreted in law. It must also be interpreted by you, because you are the person who is required to install water fittings in accordance with the law.

Many words are open to interpretation in different ways so it is of little use writing legal documents if we have one meaning in law and another meaning in practice.

Let's look at an example:

'House' in most people's language will mean a building in which people live. It is a dwelling place. But first thought conjures up a building with an upstairs and a downstairs and it may to some, seem different to a bungalow or a flat or maisonette.

However, in the terms of the Regulations, a 'house' has a particular meaning. If we do not know the correct meaning of the term we cannot correctly interpret the meaning of the Regulations.

Section 219 of the Water Industry Act 1991 gives the following definition:

'house' *means any building or part of a building which is occupied as a dwelling-house, whether or not a private dwelling-house, or which, if unoccupied, is likely to be so occupied.*

Does that match up with your idea of a house? Whether it does or not, where the Regulations or schedules under the Regulations mention a house, it is this definition or interpretation that is referred to.

As we have said, definitions have been included in a number of places in the Regulations and occasionally you may need to check on their meaning as you work through this package.

A number of terms that apply to the Regulations are defined in other legislation such as the Water Industry Act 1991 and the Building Regulations. This of course makes life more difficult for us, because not many of us are in possession of all the relevant documents, nor would we know where to find the definitions even if we had the right documents.

We have therefore attempted to put many of the definitions and interpretations together into this one document for ease of reference as you study.

We hope this will be helpful to you in your studies.

Alphabetical Index

of terms described in this document

The following is a selection of definitions and interpretations extracted from

the Water Industry Act 1991

the **"Director"** means the Director General of Water Services;   *(Para 218)*

"Food production purposes" means the manufacturing, processing, preserving or marketing purposes with respect to food or drink for which water is supplied to food production premises may be used, and for the purposes of this definition, "food production premises" means premises used for the purposes of a business of preparing food or drink for consumption otherwise than on the premises;   *(Para 93)*

"house" means any building or part of a building which is occupied as a dwelling-house, whether or not a private dwelling-house, or which, if unoccupied, is likely to be so occupied.   *(Para 219)*

"house" will include a house, bungalow, flat or similar dwelling

"local authority" means the council of a district or of a London borough or the Common Council of the City of London;   *(Para 219)*

In other words your local district or town council

"private supply" means.... a supply of water provided otherwise than by a water undertaker.....;   *(Para 93)*

"Private supply" will include any supply of water not supplied by a water undertaker, irrespective of its source.

The Regulations apply only to water supplied by a water undertaker.

"water fittings" includes pipes (other than water mains), taps, cocks, valves, ferrules, meters, cisterns baths, water closets, soil pans and other similar apparatus used in connection with the supply and use of water.
                    (Para 93)

Within the Regulations, 'water fittings' is a general term given to a variety of separate fittings and includes pipes, taps, valves, meters, cisterns and cylinders, baths, water closets and other sanitary appliances, boilers and hot store vessels, washing machines, etc.

In fact, if any item within premises contains or uses water supplied by the water undertaker, it is considered to be a water fitting.

"water for domestic purposes" refers to water used for …drinking, washing, cooking, central heating, and sanitary purposes …in a house or a building used mainly as a house. It also includes …the washing of vehicles and the watering of gardens …providing the …water is drawn from a tap within the house and without the use of a hosepipe.   *(Para 218)*

"service pipe" means …so much of a pipe which is, or is to be, connected with a water main for supplying water from that main to any premises as is to be subjected to water pressure from that main, or would be so subject but for the closing of some valve…
(Para 219)

"water main" means any pipe …which is used by a water undertaker for the purpose of making a general supply of water available to customers or potential customers …
(Para 219)

This is the main in the street from which a service pipe is connected to serve a house or other premises.

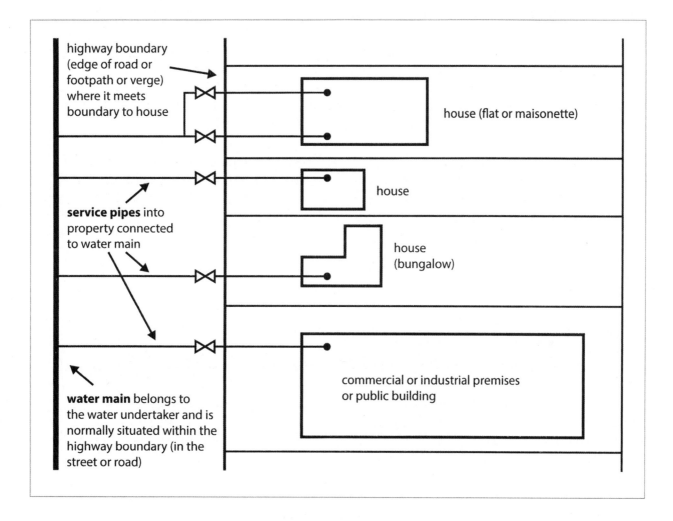

The following is a selection of definitions and interpretations extracted from

the Water Supply (Water Fittings) Regulations 1999

"the Act" *means the Water Industry Act 1991;*

"approved contractor" *means a person who:*

(a) has been approved by the water undertaker for the area where a water fitting is installed or used, or

(b) has been certified as an approved contractor by an organisation specified in writing by the regulator;

Reference is made to 'approved contractors' in Regulations 5, 6 and 7, where the Water Supply Industry is encouraged to set up Approved Contractors Schemes.

"the Directive" *means Council Directive 89/106/EEC on the approximation of laws, regulations and the administrative provisions of the member States relating to construction Products (b);*

"The Directive" (Council directive 89/106/EEC etc) refers to 'Construction Products Directive (a)' which is an instruction that, under European law, requires a product (pipe fitting etc) to be fit for its intended use. If a product meets all the essential specifications under this European law, manufacturers are entitled to use the EC mark and Member States are obliged to permit its use providing it meets with all the relevant standards and of the country in which the product is used.

"EEA Agreement" *means the agreement on the European Economic Area signed at Oporto on 2nd May 1992(c) as adjusted by the Protocol signed at Brussels 17th March 1993(d);*

"EEA State" *means a State which is a contracting party to the EEA Agreement;*

In simple terms, there are the countries in the European Community plus Norway, Iceland and Liechtenstein.

"European technical approval" *means a favourable technical assessment of the fitness for use of a construction product for an intended use, issued for the purpose of the Directive by a body authorised by an EEA State to issue European technical approvals for those purposes and notified by that State to the European Commission;*

'European technical approval' means a favourable technical assessment of the fitness for use of a construction product, issued by a body such as the British Standards Institute which is authorised by the European Commission to issue European technical approvals.

"fluid category" *means a category of fluid described in Schedule 1 to these regulations; (see page 7)*

"harmonised standard" means a standard established as mentioned in the Directive by the European standards organisation on the basis of a mandate given by the Commission of the European Economic Community and published by the Commission in the Official Journal of the European Communities;

A 'harmonised standard' is a standard produced by representatives of the various European countries within the European Standards Organisation. The standard being written and agreed, first by committee, then ratified by the Commission (Parliament) of the European Economic Community.

"material change of use" means a change in the purpose for which, or the circumstances in which, premises are used, such that after that change the premises are used (where previously they were not soused):

i) *as a dwelling;*

ii) *as an institution;*

iii) *as a public building; or*

iv) *for the purpose of the storage or use of substances which if mixed with water result in a fluid which is classified as either fluid category 4 or 5;*

'Material change of use' means a change in the purpose for which, or the circumstances in which, premises are used. For example a house (dwelling) that is turned into a nursing home (institution) or a church (public building) that is turned into a dwelling. In both cases there are changes in use that affect the requirements of the Regulations.

Under **'material change of use'** four types of premises are listed (but not defined). Definitions for these are given below:

i) **a dwelling** is a place where people live e.g. house, flat, bungalow etc.

ii) **an institution** is a building that provides living accommodation for, or for the treatment or care of, people suffering from illness or disability or those who are unable to care for themselves. Examples include certain hospitals, schools, homes for the young or old, but not day centres.

iii) **a public building** can be described as premises designed and built for use by the general public including such buildings as theatres, schools or colleges of education, public libraries, halls where people meet, and places of worship.

iv) **buildings used for the storage or use of substances of fluid category 4 or 5.** This can include any type of building providing the said substances are stored.

Note: the above do not give definitive lists of examples.

"regulator" means:

(a) *in relation to any water undertakers whose area of appointment is wholly or mainly in Wales and their area of appointment, The National Assembly for Wales;*

(b) *in relation to all other water undertakers and their area of appointment, the Secretary of State;*

"supply pipe" means so much of any service pipe that is not vested in the undertaker; and Paragraph 1 of Schedule 2 has effect for the purposes of that Schedule. (See also page 11)

The following selection of definitions and interpretations is extracted from:

Schedule 1 of the Water Supply (Water Fittings) Regulations 1999

Fluid categories

Fluid category 1

Wholesome water supplied by a water undertaker and complying with the requirements of regulations made under Section 67 of the Water Industry Act 1991(a)

Fluid category 2

Water in fluid category 1 whose aesthetic quality is impaired owing to:

(a) a change in its temperature, or

(b) the presence of substances or organisms causing a change in its taste, odour or appearance including water in a hot water distributing system

Fluid category 3

Fluid which represents a slight health hazard because of the concentration of substances of low toxicity, including any fluid which contains:

(a) ethylene glycol, copper sulphate solution or similar chemical additives, or

(b) sodium hypochlorite (chloros and common disinfectants)

Fluid category 4

Fluid which represents a significant health hazard because of the concentration of toxic substances, including any fluid which contains:

(a) chemical, carcinogenic substances or pesticides (including insecticides and herbicides), or

(b) environmental organisms of potential health significance.

Fluid category 5

Fluid representing a serious health hazard because of the concentration of pathogenic organisms, radio active or very toxic substances. including any fluid which contains:

(a) faecal matter or other human waste;

(b) butchery or other animal waste; or

(c) pathogens from any other source.

Note: A comprehensive list of examples is shown in Tables 6.1 on pages 10 to 12 of Manual 8.

The following selection of definitions and interpretations is extracted from:

Schedule 2 of the Water Supply (Water Fittings) Regulations 1999

*"**backflow**" means movement of fluid from downstream to upstream within an installation due to:*

(a) **backpressure,** *that is pressure generated at any point in a system which is greater than the pressure upstream of that point at the same elevation, or*

(b) **backsiphonage,** *that is pressure generated at any point in a system which is greater than the pressure upstream of that point at the same elevation;*

> Backflow can be said to be flow in a direction contrary to the normal intended direction of flow

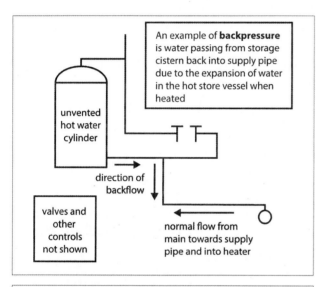

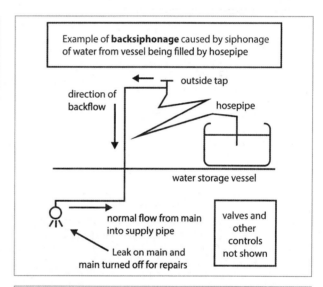

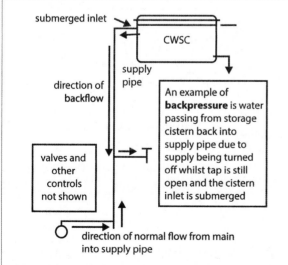

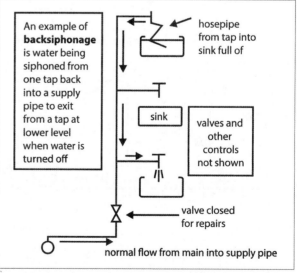

*"**contamination**" includes any reduction in chemical or biological quality of water due to raising its temperature or the introduction of polluting substances;*

*"**cistern**" means a fixed container for holding water at atmospheric pressure;*

*"**storage cistern**" means a cistern for storing water for subsequent use, not being a flushing cistern.*

Cistern can include a feed and expansion cistern, feed cistern, or storage and feed cistern

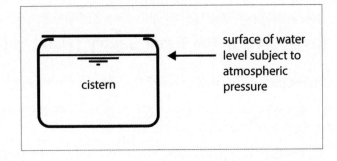

*"**flush pipe**" means a pipe conveying water from a flushing cistern or flushing valve to a water closet pan or urinal;*

*"**flushing cistern**" means a cistern provided with a valve or device for controlling the discharge of the stored water into a water closet pan or urinal;*

*"**flushing trough**" means a flushing apparatus which combines several discharging units into a single cistern to allow frequent flushing of two or more water closet pans;*

*"**spillover level**" means the level at which the water in a cistern or sanitary appliance will first spill over if the inflow of water exceeds the outflow through any outflow pipe and any overflow pipe;*

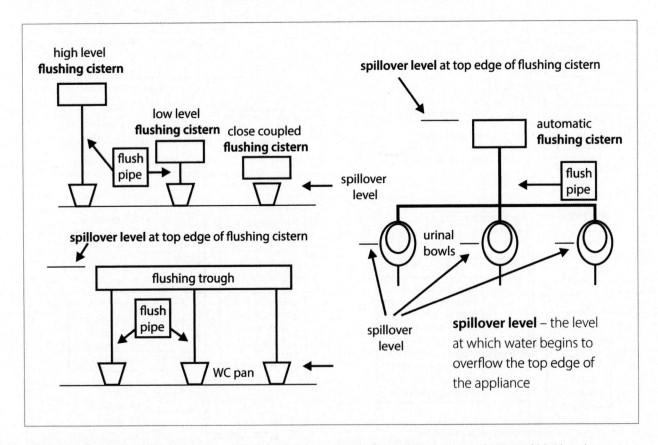

*"**pressure flushing cistern**" means a WC flushing device that utilises the pressure of water within the cistern supply pipe to compress air and increase the pressure of water available for flushing a WC pan;*

*"**pressure flushing valve**" means a self-closing valve supplied with water directly from a supply pipe or a distributing pipe which when activated will discharge a pre-determined flush volume;*

"combined feed and expansion cistern" means a cistern for supplying cold water to a hot water system without a separate expansion cistern;

"expansion cistern" means a cistern connected to a water heating system which accommodates the increase in volume of that water in the system when the water is heated from cold;

"primary circuit" means an assembly of water fittings in which water circulates between a boiler or other source of heat and a primary heat exchanger inside a hot water storage vessel;

"secondary circuit" means an assembly of water fittings in which water circulates in supply pipes or distributing pipes to an from a hot water storage system;

"secondary system" means that part of any hot water system comprising the cold feed pipe, any hot water storage vessel, water heater and pipework from which hot water is conveyed to all points of draw-off;

"vent pipe" means a pipe open to the atmosphere which exposes the system to atmospheric pressure at its boundary;

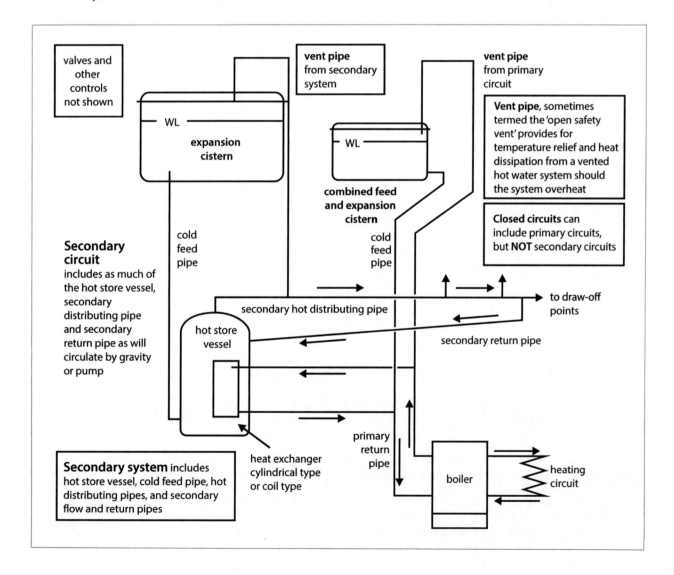

"overflow pipe" *means a pipe from a cistern in which water flows only when the water level in the cistern exceeds its normal maximum level;*

"warning pipe" *means an overflow pipe whose outlet is located in a position where the discharge of water can be readily seen.*

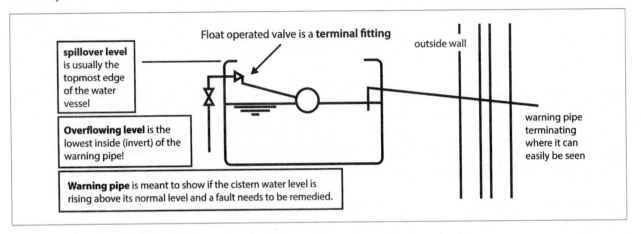

"distributing pipe" *means any pipe (other than a warning, overflow or flush pipe) conveying water from a storage cistern, or from hot water apparatus supplied from a cistern and under pressure from that cistern;*

"supply pipe" *means so much of any service pipe as is not vested in the water undertaker;*

"servicing valve" *means a valve for shutting off the flow of water in a pipe connected to a water fitting for the purpose of maintenance or service;*

"stopvalve" *means a valve, other than a servicing valve, for shutting off the flow of water in a pipe;*

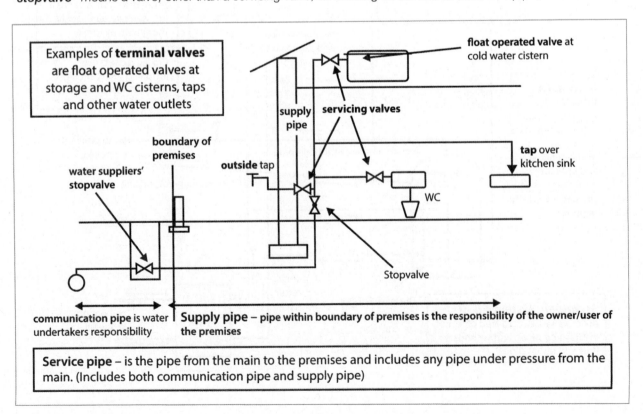

"upstand" *means a pipe arrangement consisting of:*

(a) *a branch pipe serving an appliance, where the height of connection to a supply pipe or distributing pipe is not less than a specified distance above the spillover level of an appliance, or*

(b) *a pipe surmounted by an anti-vacuum valve whose outlet is located not less than a specified distance above the spill-over level of an appliance.*

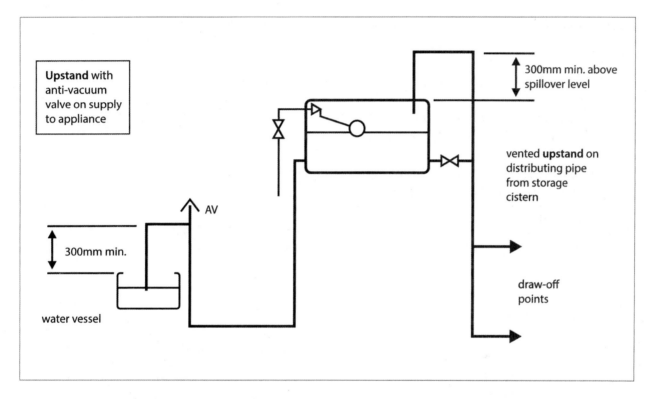

"concealed water fitting" *means a water fitting that:*

(a) *is installed below ground;*

(b) *passes through or under any wall, footing or foundation;*

(c) *is embedded in any wall or solid floor;*

(d) *is enclosed in any chase or duct, or*

(e) *is in any other position which is inaccessible or renders access difficult;*

Inaccessible positions could include inside at the back of kitchen units, or roof spaces without proper means of access etc.

"surge" *means a rapid increase of pressure in a pipeline;*

A common cause of surge is the initial operation of a pump.

"terminal fitting" *means a water outlet device.*

"instantaneous water heater" *means an appliance in which water is immediately heated as it passes through the heater;*

"unvented hot water storage vessel" *means a hot water storage vessel that is not provided with a vent pipe but is fitted with safety devices to control primary flow, prevent backflow, control working pressure and accommodate expansion;*

"pressure relief valve" *means a pressure-activated valve which opens automatically at a specified pressure to discharge fluid;*

"temperature relief valve" *means a valve which opens automatically at a specified temperature to discharge water;*

"combined temperature and pressure relief valve" *means a valve capable of performing the function of both a temperature relief valve and a pressure relief valve;*

"operational discharge" *means a discharge of water resulting from the operation of fittings, equipment or appliances which is necessary for their operation;*

'operational discharge' could include the running to waste of water from an unvented hot water storage vessel where water is permitted to discharge through the operation of a temperature relief valve or a pressure (expansion) relief valve.

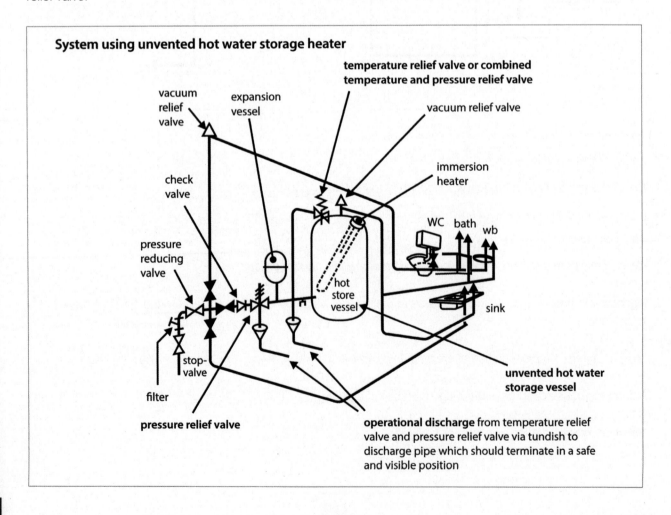

System using unvented hot water storage heater

Colour code identification of new water services requirements

Typical application	Basic colour identification band				Basic colour identification band
Source of water Potable designation for water meeting drinking water standards					Water provided from the public supply (i.e. water undertaker)
					Water derived from a source other than the public supply (i.e. private borehole, well etc.)
End use water quality An additional black band to be applied where the end use fluid is not intended to meet standards for drinking water					Public water supply system
					Any other water source
Safety systems Fire systems connected to a drinking water mains and containing no additives, following an assessment, may be considered for potable designation					Public water supply system
					Any other source
Non-potable designation to be applied to fire systems which, are fed from a dedicated fire storage cistern, containing additives or where there is doubt regarding the water quality					Public water supply system
					Any other source

Colour coding key

Green   ▬   Black   ▬   Flint Grey   ▬   Red   ▬   Auxiliary Blue   ▬